Latex Bonding Chemistry and Processes

LATEX BONDING CHEMISTRY AND PROCESSES

Edited by:

Dr. Saeed Farrokhpay

ARCLER PRESS

www.arclerpress.com

Latex Bonding Chemistry and Processes

Dr. Saeed Farrokhpay

Arcler Press
224 Shoreacres Road
Burlington, ON L7L 2H2
Canada
www.arclerpress.com
Email: orders@arclereducation.com

ISBN: 978-1-77469-824-2 (Hardcover)

ABOUT THE EDITOR

Dr Saeed Farrokhpay is a Chemical Engineer with several years of experience in mineral & material processing. He obtained his PhD from University of South Australia in 2005. He is currently a Technical Consultant in Australia. He has worked for more than 20 years at mineral and chemical industries, universities and research centers around the world. Dr Farrokhpay has published more than 90 papers in high ranked journals and conference proceedings. He has also edited several technical and scientific books, and served as an editorial board member of several international scientific journals.

TABLE OF CONTENTS

LIST OF FIGURES

LIST OF TABLES

LIST OF ABBREVIATIONS

ABS	Acrylonitrile-Butadiene-Styrene
AFM	Atomic Force Microscopy
AIBN	Azobisisobutyronitrile
AMBN	2,2'-Azobis(2-Methylpropionitrile)
ATRP	Atom Transfer Radical Polymerization
BPO	Benzoyl Peroxide
CMC	Critical Micelle Concentration
CO_2	Carbon Dioxide
CPC	Cetylpyridinium Chloride
CPVC	Critical Pigment Volume Concentration
CSTRs	Continuously Stirred Tank Reactors
CTAB	Cetyltrimethylammonium Bromide
DLS	Dynamic Light Scattering
DMA	Dynamic Mechanical Analysis
DOSY	Diffusion-Ordered Spectroscopy
DSC	Differential Scanning Calorimetry
EVA	Ethylene-Vinyl Acetate
EVCL	Ethylene/Vinyl Chloride
FSC	Forest Stewardship Council
FTIR	Fourier Transform Infrared
GC	Gas Chromatography
GPC	Gel Permeation Chromatography
GPO	Group Purchasing Organization
HDC	Hydrodynamic Chromatography
HEC	Hydroxyl Ethyl Cellulose
HIPS	High Impact Polystyrene
HLB	Hydrophilic/Lipophilic Balance

ISO	International Standards Organization
IUPAC	International Union of Pure and Applied Chemistry
LCA	Life Cycle Assessment
MALDI MS	Matrix-Assisted Laser Desorption Ionization Mass Spectrometry
MFFT	Minimum Film-Forming Temperature
Mw	Molecular Weight
NMR	Nuclear Magnetic Resonance
NR	Natural Rubber
NRL	Natural Rubber Latex
NTA	Nanoparticle Tracking Analysis
OM	Optical Microscopy
PDI	Polydispersity Index
PFR	Plug Flow Reactor
PSA	Pressure-Sensitive Adhesive
PU	Polyurethane
PVA	Polyvinyl Alcohol
PVC	Polyvinyl Chloride
QELS	Quasielastic Light Scattering
RAFT	Reversible Addition-Fragmentation Transfer
RSPO	Roundtable on Sustainable Palm Oil
SBR	Styrene-Butadiene Rubber
SBS	Styrene-Butadiene-Styrene
SDS	Sodium Dodecyl Sulfate
SEBS	Styrene-Ethylenebutadiene Styrene
SEC	Size Exclusion Chromatography
SEM	Scanning Electron Microscopy
SFRP	Stable Free Radical Polymerization
SIS	Styrene-Isoprene-Styrene
SLS	Sodium Lauryl Sulfate
SLS	Static Light Scattering
STDs	Sexually Transmitted Diseases
TEM	Transmission Electron Microscopy
Tg	Transition Temperature
ToF-SIMS	Time-of-Flight Secondary Ion Mass Spectrometry

UV	Ultraviolet
VA	Vinyl Acetate
VAE	Vinyl Acetate-Ethylene
VOCs	Volatile Organic Compounds
XPS	X-ray Photoelectron Spectroscopy

PREFACE

Latex bonding is a vital process in the world of chemistry and materials science. It involves the creation of a strong adhesive bond between latex particles, which are typically made of polymers such as styrene-butadiene or acrylics. Latex bonding is commonly used in industries such as paper manufacturing, textile production, and construction. The process typically begins with the dispersion of latex particles in water, forming a stable colloidal suspension. Then, various additives such as surfactants, coagulants, and crosslinking agents are introduced to promote bonding and stabilize the latex. The latex is then applied to the desired substrate and undergoes a curing process, during which the water evaporates, and the latex particles coalesce to form a continuous and durable film. Latex bonding chemistry and processes play a crucial role in creating high-performance materials with excellent adhesive properties and are continuously researched and optimized to meet the diverse demands of modern industries.

This book is structured into eight insightful chapters, each offering a comprehensive exploration of a key dimension in the field of latex bonding. This book provides a holistic understanding of the subject matter, covering both theoretical foundations and practical applications. Let's delve into each chapter:

Chapter 1 serves as an introduction to latex materials, providing an overview of their composition, properties, and diverse applications across industries. Chapter 2 delves into the fundamentals of latex chemistry, elucidating the chemical structure and behavior of latex particles.

Chapter 3 focuses on the physics and mechanics underlying latex bonding. It discusses concepts such as particle coagulation, film formation, and the role of intermolecular forces in adhesive strength. Chapter 4 delves into the preparation and characterization of latex. It covers methods for synthesizing latex particles, including emulsion polymerization and suspension polymerization. Chapter 5 addresses the growing importance of environmentally friendly practices in latex production and disposal. It discusses sustainable approaches to latex synthesis, waste management strategies, and the development of biodegradable latex materials.

Chapter 6 focuses on the influence of additives on latex bonding. It explores the role of surfactants, coagulants, crosslinking agents, and other additives in enhancing bonding performance. Chapter 7 highlights the wide-ranging applications of latexes. It showcases their utilization in fields such as paints, adhesives, coatings, textiles, and healthcare products. Chapter 8 takes a forward-looking approach, discussing future trends and troubleshooting in latex bonding. It examines emerging technologies, advancements in formulation strategies, and potential challenges that may arise.

This handbook is a collaborative effort of experts in the field of latex bonding chemistry, combining their collective knowledge, research, and practical expertise. The comprehensive nature of each chapter, coupled with real-world examples and case studies, ensures that readers gain a thorough understanding of latex bonding chemistry and its practical implications. It is our hope that this handbook serves as an invaluable resource, inspiring further exploration, innovation, and excellence in the field of latex bonding.

—***Author***

CHAPTER

1 Fundamentals of Latex Materials

Contents

1.1. INTRODUCTION

Latex can be classified as a type of colloidal dispersion. The composition of the substance in question comprises of polymeric particles, typically measuring only a few hundred nanometers in diameter, that are dispersed within an aqueous medium. The dispersion is typically composed of particles that constitute approximately 50% of its total weight. The presence of a diverse combination of pigments, surfactants, plasticizing agents, and rheological modifiers may be observed in various applications. The entire dispersion exhibits colloidal stability, indicating its ability to maintain dispersion without the occurrence of particle sedimentation or formation of residual sludge at the bottom, even over extended periods of time. The term 'latex' will be employed as a concise substitute for a moist dispersion in the present publication. On occasion, the term "latex" is employed as an adjective, such as in the phrase "latex film." It is important to note that the correct plural form of the term 'latex' is 'latices.' This should not be confused with the term 'lattices.' (An alternative term that may be used is "latexes;" however, for the purposes of this book, the term "latices" will be utilized) (Keddie & Routh, 2010).

Did You Know?

Latex instability and coagulation can occur due to mechanisms like flocculation and Ostwald ripening.

While colloids have a historical usage dating back to ancient civilizations such as Egypt and Greece, the scientific study of colloidal dispersions did not commence until the nineteenth century with the introduction of the term "colloid" by Thomas Graham. The term is of Greek origin, specifically derived from the word "coll" which denotes "glue." Robert Brown made a significant contribution to colloid science by documenting the random motion of pollen particles when spread in a fluid. The current scientific terminology for this occurrence is Brownian motion, which serves as a distinctive feature of a colloidal dispersion. The observed phenomenon can be attributed to the unequal distribution of forces among the colloidal particles as well as the solvent molecules, thus serving as a proof of the presence of molecules. A dispersion that exhibits Brownian motion is commonly referred to as colloidal. According to the International Union of Pure and Applied Chemistry (IUPAC), colloidal particles can be defined as having a minimum of one dimension that falls within the range of 1 nm to 1 μm (Ho, 2021).

Figure 1.1. An emulsion polymerization reactor. The structure spans an entire building and covers two floors. The top of the reactor houses the stirring motor.

Source: Akzo Nobel, Creative Commons License.

1.2. LATEX SYNTHESIS AND USES

Emulsion polymerization is the prevalent method for producing latex in both industrial and laboratory settings. The scalability of the process is exemplified by the sizable reactors depicted in Figure 1.1. The top 2 photographs serve as an indication of the scale of the two-story structure. The photograph presented at the bottom depicts the uppermost section of the reactor alongside the motor responsible for the stirring process. The process of emulsion polymerization and its recent advancements, such as miniemulsion polymerization, have been extensively documented in various sources, as reported by Lovell and El-Aasser in 1997. An examination of the process of polymerization is outside the purview of our current discussion (Bar et al., 2009).

Latex can be produced through the polymerization of various monomers, including but not limited to acrylates (such as methyl methacrylate, butyl acrylate, and ethylhexyl acrylate), butadiene, vinyl acetate, and styrene. Copolymers are widely utilized in various applications. The glass transition temperature (Tg) of a copolymer is determined by the molar ratio of its constituent monomers, as this ratio enables long-range motion of the polymer molecule's backbone, resulting in a transition from a solid to a liquid-like state. The viscosity of a polymer exhibits a sub-exponential decline upon surpassing the Tg threshold with increasing temperature. It is imperative that the temperature utilized in the formation of a latex film exceed the Tg of the (co)polymer (McDonald & Devon, 2002).

Figure 1.2. Latex binders market global forecast to 2026.

Source: Markets and markets, Creative Commons License.

The utilization of latex in contemporary waterborne paints is advantageous for individuals who engage in the coating of residential structures. The predominant application of latex in households pertains to architectural along with decorative paints, that are commonly composed of copolymers of acrylic or styrene-acrylic. The majority of readers are likely to possess knowledge regarding commercial latex paints, as exemplified in Figure 1.3 (Charleux et al., 2010).

Typically, household varnishes and paints are applied to limited surface areas and subsequently left to dry. The regulation of viscosity holds significant importance in the context of paint formulations. At low shear rates, such as when applied in a vertical orientation on a wall, it is expected that the paint will exhibit high viscosity and resist flow or dripping. At elevated shear rates, it is desirable for the viscosity to be reduced to facilitate uniform dispersion when applying with a brush.

Figure 1.3. Latex paint is marketed in the UK and other countries under the trade name of Dulux, as seen in this advertising material, showing a well-recognized mascot.

Source: Akzo Nobel, Creative Commons License.

Therefore, rheology modifiers play a crucial role as a constituent component in the formulation of commercial paints. The desirable property of shear thinning behavior, wherein the viscosity of a substance decreases just like the shear rate rises, is crucial in ensuring optimal spreading when using a brush and applicator. The customary thickness of paint films in household applications is approximately one hundred micrometers. Table 1.1 displays the standard compositions for three categories of commercial emulsion (i.e., latex) paint, while additional formulations can be found in literature sources (Heldmann et al. 1999). Glossy paints exhibit a high degree of light reflectance, while matte paints demonstrate a low level of reflectivity. On the other hand, silk paints possess an intermediate level of reflectivity (Fujii et al., 2009; Rajkuberan et al., 2017).

It is noteworthy that the paint formulations contain no more than 30 % by weight of the latex polymer. It should be noted that there are rheology modifiers as well as defoamers present. In order to achieve complete coverage of hydrophobic surfaces, wetting agents are frequently incorporated. Matte paints are often supplemented with extenders, which are low-cost minerals utilized as fillers, to increase the coverage of the more costly polymers.

Extenders are avoided in gloss paints due to their tendency to cause light scattering (Hudlikar et al., 2012).

Latex is progressively gaining popularity as a binding agent in industrial coatings that are utilized in factory production, as well as in diverse forms of protective finishes for vehicles, aircraft, ships, appliances, and other applications. The requirements placed on industrial as well as protective coatings are stringent and extend beyond the mere enhancement of an object's visual appeal. The coatings must exhibit resistance to corrosion, abrasion, wear, fire, and chemical attack. In addition to its application in paints and coatings, latex finds utility in a diverse range of products such as pressure-sensitive adhesives (PSAs), sealants, carpet backings, constructing ingredients (including cements as well as mortars), paper coatings, inks, as well as latex gloves and condoms. Each of these will be briefly considered in this context (Smeets et al., 2017).

Pressure-sensitive adhesives (PSAs) exhibit a permanent tackiness at ambient temperature and offer prompt and enduring adhesion to a wide range of surfaces. Tapes as well as sticky labels are widely recognized for their practical use in various applications. PSAs have numerous distinct applications in both household settings (as shown in Figure 1.4) and automotive contexts (as depicted in Figure 1.5). The copolymers are commonly synthesized using monomers like butyl acrylate or 2-ethyl hexyl acrylate to lower their glass transition temperature, which typically falls in the vicinity of -50°C (Guidelli et al., 2011).

In an industrial setting, the process of depositing films involves the application of a thin layer onto elongated bundles of substrate material that are propelled at elevated velocities through specialized coaters. The transfer coating technique is commonly employed for the production of labels. A sandwich configuration, as depicted in Figure 1.6, has been fabricated. The process involves the application of latex onto a release liner that possesses a surface with low energy, typically composed of a silicone polymer, followed by the formation of a film. The acceleration of the drying process can be achieved by subjecting the film, which is affixed to the backing layer, to an extended oven or by exposing it to a stream of heated air. The customary dry layer thickness for various applications is generally 20 micrometers. Subsequently, a "face material" or facestock is applied onto the latex binding surface through pressure. The facial exterior of the facestock is suitably embellished to cater to the adhesive's intended use. In order to apply the adhesive, one must first remove it from the releasing liner and subsequently affix it onto the desired surface, like a beer bottle. It is imperative that the ultimate cinematic product exhibits homogeneity and is devoid of any imperfections, like pinholes or surface undulations (Butler et al., 2005; Das et al., 2011).

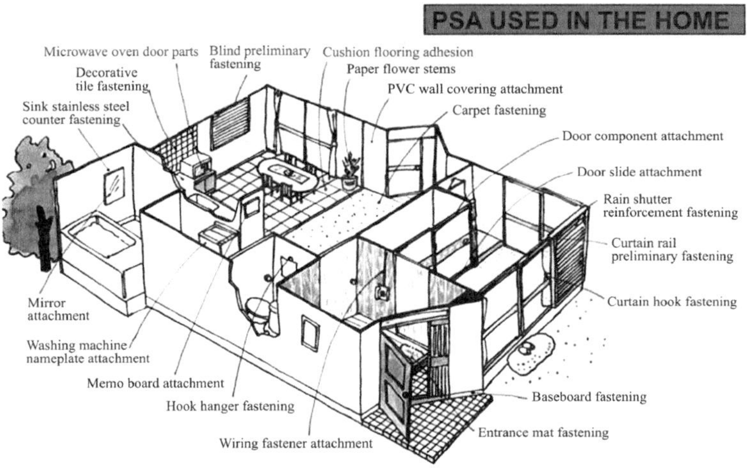

Figure 1.4. Uses of pressure sensitive adhesives in the home.

Source: Cytec specialty chemicals, Creative Commons License.

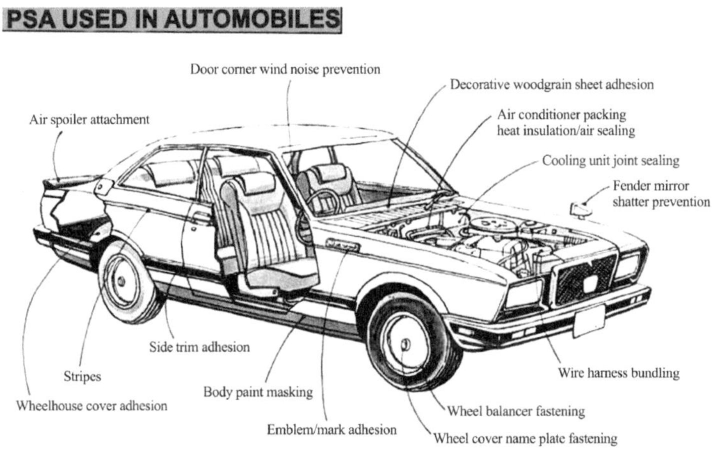

Figure 1.5. Uses of pressure sensitive adhesives in a car.

Source: Cytec specialty chemicals, Creative Commons License.

In a conventional industrial coating system, such as the one depicted in Figure 1.6, a steel cylinder is revolved through a pool of moist latex, collecting the substance as it rotates. Subsequently, the backing coat is rapidly applied onto the wet roll's rotation to create a substantial coating. The superfluous latex present in the film is removed through the utilization of a grooved bar, commonly referred to as a 'Meyer rod,' which is passed over the surface (Bano et al., 2018).

An alternative technique, known as 'gravure,' employs a cylinder that features etched grooves. Upon immersion of the gravure roll in a latex reservoir, the grooves of the roll acquire a certain amount of latex. Subsequently, the roll is brought into interaction with a backing layer, resulting in the deposition of a moist film onto said layer. The gravure process is capable of achieving higher coating speeds in comparison to the majority of other processes. Coating speeds may vary between 100 to more than 1000 meters per minute, contingent upon the specific process employed (Mondal et al., 2011).

Acrylic polymers are commonly utilized in the production of textiles and carpet backings. According to Campos et al. (2006), the application of a coating on the material surface enhances its durability and confers resistance to various solvents (Thakore et al., 2019).

Hydrophobic polymer latices are frequently incorporated into construction materials. According to Stancato et al. (2005), incorporating these elements into the material can decrease its water permeability, ultimately resulting in a longer lifespan due to decreased corrosion (Juhue & Lang, 1995).

Latex is utilized as a binding agent in paper coatings to facilitate the integration of inorganic fillers and clays. According to Zang and Aspler (1995) and Alturaif et al. (1995), the utilization of latex coatings on paper is a cost-effective alternative to wood pulp. Additionally, this method facilitates enhanced ink penetration during the printing process (Teo et al., 2009).

According to Hutton and Parker (2008), latex can be utilized as a binding agent for pigment and dye particles in the inks intended for printing upon paper as well as card. The substitution of solvent-based substances with water-based inks has garnered increasing attention. Colloidal dispersions are commonly utilized inks in inkjet printing, as per Calvert's research in 2001. Wong et al. (1988) have also exhibited the application of latex particles in an inkjet printer (Grunlan et al., 2001).

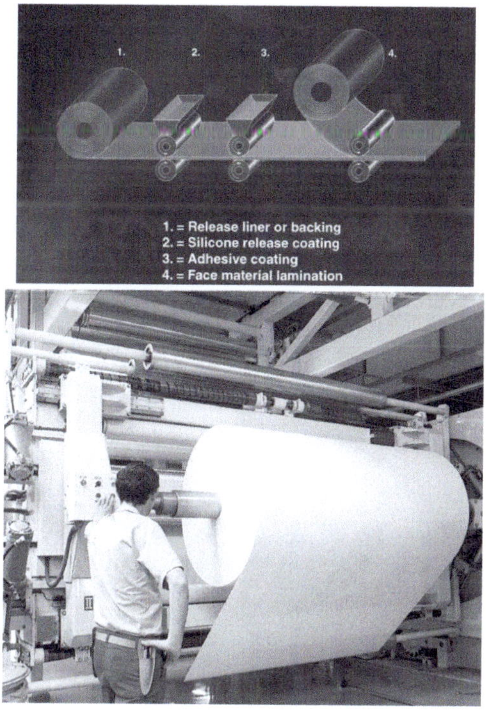

Figure 1.6. Industrial coating of pressure-sensitive adhesive films. A in the transfer process, the latex adhesive 3 is deposited onto a silicone release coating 2 on a release liner 1. Then the face material 4 is laminated onto the adhesive surface.

Source: Chemical Concepts, Creative Commons License.

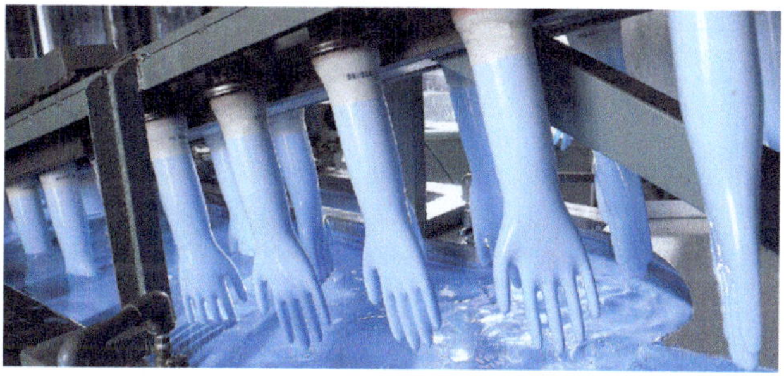

Figure 1.7. Latex gloves are being made by dipping molds into a natural latex dispersion and with drawing.

Source: Riverstone holdings limited, Creative Commons License.

The production process of latex gloves involves immersing a ceramic mold resembling a human hand in a latex dispersion as well as subsequently drying it, as depicted in Figure 1.7. Condoms are produced using a similar manufacturing process, albeit with a distinct mold (Barkade et al., 2013).

1.3. HISTORICAL CONTEXT AND ECONOMIC IMPORTANCE

A naturally occurring form of latex can be obtained from the Hevea Brasiliensis tree, which has its origins in Brazil, as indicated by its Latin nomenclature (Fig. 1.8). During the twentieth century, various Asian nations established plantations to facilitate the expansion of the rubber industry. The composition of natural latex primarily consists of cis-isoprene that is emulsified through a protein. The disruption of latex supply from Malaysia during World War II led to shortages, which in turn prompted the advancement of synthetic latex produced through emulsion polymerization. In 1948, the Glidden Company introduced the initial commercially accessible latex. As a result, a substantial synthetic latex sector expanded throughout the remaining period of the century. Another significant aspect to consider is the prevalence of protein allergy in natural latex, with approximately 10% of individuals exhibiting sensitization, as reported by Turjanmaa et al. (1996) and Liss et al. (1997). The aforementioned limitation of natural latex has stimulated the expansion of the synthetic latex industry as a viable substitute, particularly in the domain of gloves (Homma, 1992; Keddie et al., 2010).

Figure 1.8. The milky substance that is observed to drip from the Hevea Brasiliensis tree is identified as sap, and serves as the organic source of latex. The liquid latex is being accumulated within a receptacle situated at the base of the tree, with the intention.

Source: Daniel, Creative Commons License.

The global market for latex materials exhibits a significant expansion trend, as evidenced by its substantial sales figures of 17.9 billion US dollars in 2007. The extensive utilization of polymeric coatings across various industries renders the economic impact of the coatings sector notably significant. The paper manufacturing sector in the European Union produces approximately ninety-one million tons of paper each year, utilizing coatings to achieve a lustrous white appearance. According to Urban and Takamura's (2002) research, the European Union's yearly market for paper coverings amounts to 800 million Euros. Adhesives as well as sealants are utilized across a wide range of over 100 end-use product markets. In the European region, the demand for these materials amounted to four million tons in 2008. A projected increase of 2.5% is anticipated during the period spanning from 2008 to 2011. According to von Dungen (2009), water-based adhesives that are derived from latex accounted for 37% of the 11.3 billion Euro (15.8 billion United States dollars) European adhesive market. The aqueous-based industry is expanding while displacing other manufacturing procedures that discharge organic solvents (Peluso, 2012).

According to Tan's (2008) report, the production of latex gloves amounted to 135 billion in the year 2008, with an annual growth rate of approximately 10%. According to Tan's (2008) report, the prevalence of natural latex allergies has led to a rise in demand for nitrile latex gloves in the United States. In 2007, the market share of nitrile latex gloves increased to 26.8%, while the market share of natural rubber gloves decreased from 48% to 41%. Malaysia holds the top position as the largest exporter of rubber gloves. According to available data, the aggregate quantity of glove pairs exported in 2007 amounted to 41.7 billion pairs, which were valued at 5.9 billion Malaysian dollars (equivalent to 1.72 billion US dollars). This figure is equivalent to approximately 1% of Malaysia's Gross Domestic Product, as reported by the Department of Statistics in 2008. According to Urban and Takamura's (2002) findings, the yearly utilization of polymer for textiles amounted to 160,000 dry tons in 2001, while the consumption of latex for carpet backings reached 500,000 dry tons in 1999 (Keddie & Routh, 2010).

Latex is commonly linked with aqueous coatings utilized for the painting of both interior and exterior surfaces of edifices and residences, commonly known as 'architectural coatings.' According to projections, stringent environmental regulations will lead to a significant shift towards water-based architectural paints in Europe in 2011, with an estimated 88% of sales being attributed to such products. These paints will utilize latex polymers to be their primary binding agent. It is projected that water-based architectural paints will account for 73% of the global market share in the upcoming year. It is projected that the total output of architectural paints will attain twenty-one million metric tons by the year 2011, with an estimated worth of forty-seven billion US dollars. Notwithstanding the challenging economic conditions, the European Coatings Journal (2008) forecasts an approximate 4% yearly expansion rate worldwide during the period of 2006–2011 (Agrawal & Konno, 2009).

1.4. IMPORTANCE OF LATEX MATERIALS

The term 'latex' typically evokes associations with latex gloves. The utilization of natural rubber latex (NRL) in glove production is among the various types of latexes found in prevalent household commodities, such as latex colors, paper, floor polishes, nonwoven textiles, adhesives, and carpeting. The application of latexes to commercial products is a common practice aimed at improving targeted performance attributes such as chemical resistance, dimensional stability, and durability. Latexes are commonly referred in as polymer dispersions, polymer colloids, orlatex materials within academic literature (Gawchik, 2011).

Latexes are liquid substances that usually contain water and are composed of dispersed microscopic polymer particles. Latexes originate as emulsions of monomers, where monomer droplets are dispersed immiscibly in water. The emulsions undergo polymerization subsequently, resulting in the formation of latexes. The process of polymerization that is commonly referred to as emulsion polymerization is being discussed (Caballero & Quirce, 2015).

Although solvent-based latexes are present, their usage is restricted to a few specific applications. The utilization of organic solvents to be the continuous phase is necessary for the production of inverse emulsions. The redispersion of latex particles into organic solvents has been observed, as exemplified by the PVC plastisols discussed in Section 6. The utilization of organic solvents within latexes is forbidden due to environmental concerns. The impetus behind the creation of water-based latexes has largely stemmed from the aspiration to cease the utilization of solvent-based polymer solutions (Turjanmaa et al., 2002).

The inception of emulsion polymerization can be traced back to the late 1920s when industrial laboratories sought to create synthetic rubber latexes as a substitute for natural rubber latexes in tire production. The synthetic emulsion polymerization process gained significant traction and was extensively researched and used during World War II, owing to the limited availability of natural rubber resources. Subsequently, novel categories of latexes are being introduced to the paper, textiles coatings, and adhesives sectors for commercial purposes. At present, latexes of various kinds are being manufactured in quantities surpassing eight million dry metric tons annually (Mondal et al., 2011).

1.5. ADVANTAGES OF LATEX MATERIALS

Emulsion polymerization is characterized by the crucial aspect of flexibility. The properties of Latex can be customized to suit the specific requirements of the application. Emulsion polymerization is a highly adaptable process that allows for the utilization of diverse monomers, processing techniques, and additives. A diverse range of commodities possessing distinct characteristics can be produced. The process of emulsion polymerization facilitates the creation of particles that possess customized characteristics such as size, composition, morphology, as well as molecular weight. Incorporation of functional groups is also feasible. Various blends of distinct latex types have been developed to achieve the intended characteristics without the need for copolymerization, as reported in literature sources (Kaneko et al., 2010).

In the majority of other polymerizations, the rate of polymerization (Rp) exhibits an inverse relationship with the molecular weight. The utilization of a significant quantity of radicals to achieve an elevated rate of polymerization concurrently leads to the generation of polymer with low molecular weight. The phenomenon of radical segregation through compartmentalization within polymerizing particles can lead to a concomitant increase in both the rate of polymerization as well as the molecular weight in emulsion polymerization (Klass, 2007). Furthermore, the inherent low stickiness of latexes facilitates efficient heat transfer during the process of polymerization and enables superior substrate coating flowability. Subsequent to the addition of water, the process of evaporation can be expedited, thereby facilitating the coalescence of latex particles and the eventual formation of a cohesive polymer film. Compared to solvent-based coatings, based on water emulsion polymer coatings exhibit superior environmental friendliness. The utilization of solvent-based polymer applications is discouraged by the imposition of environmental regulations that restrict the discharge of volatile organic compounds (VOCs) (305). Latexes that are water-based have the capability to execute comparable functions to those of solvent-based systems, while avoiding any adverse environmental effects (Zin et al., 1995).

1.6. CLASSES OF LATEXES

The classes of latexes are as follows:

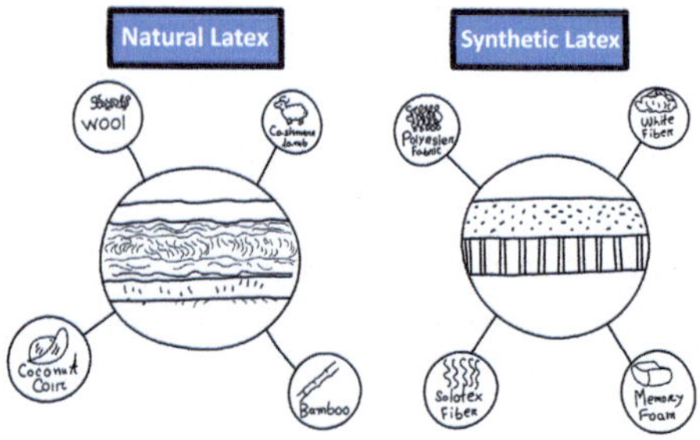

Figure 1.9. Difference between natural and synthetic latex.

Source: National Mattress Outlet Plus+, Creative Commons License.

1.6.1. Natural Latexes

The most primitive forms of latex materials are those that occur naturally. For centuries, latexes of natural rubber (NR) have been obtained from the rubber plant, Hevea Brasiliensis. Latexes consist of dispersions of large-molecular-weight linear cis-1,4-polyisoprene particles with diameters ranging from 10 nm up to several microns. The colloidal nature of natural rubber particles necessitates their separation from the aqueous phase through coagulation prior to processing. Due to its stereoregular molecular structure, natural rubber exhibits exceptional mechanical properties that remain unmatched by contemporary synthetic rubber materials (Jacob et al., 1993).

Figure 1.10. Image of natural latex harvesting.

Source: Sleep on Latex, Creative Commons License.

The material known as natural rubber is classified as an elastomer due to its inherent ability to undergo significant deformation and subsequently revert back to its initial configuration upon the cessation of stress. A material's ability to display elasticity is contingent upon its molecular weight, crosslink density, and adequate chain mobility. Elastic materials have numerous commercial applications. Rubber is commonly employed in various industrial contexts, such as for sealing purposes, as a means of mitigating shock, as a flexible barrier, as well as in scenarios where optimal traction is of utmost importance. The utilization of natural rubber has been prevalent in the production of automobile tires and latex gloves (Verhaar, 1959).

1.6.2. Synthetic Latexes

Currently, the majority of latexes produced can be categorized into three primary classes of materials. The aforementioned materials are classified as styrene-butadiene rubbers, polyacrylics, and polyvinyl acetate. The majority of latexes are generated as copolymers throughout these overarching categories, with the copolymer configuration being contingent on the intended latex characteristics. Various latex classes, such as polyurethanes, polyvinyl chloride dispersions, and acrylonitrile-butadiene rubbers are being manufactured with relatively lower commercial significance (Dillon et al., 1951).

1.6.2.1. Styrene-Butadiene Rubber

In contrast to natural rubber, synthetic polybutadiene rubbers exhibit a lack of stereoregularity in their structural composition. The preferential outcome during polymerization is the formation of linear polybutadiene chains through cis-1,4 addition. Nonetheless, trans-1,4 as well as 1,2 or 3,4 additions take place, leading to a decrease in the crystallization induced by extension and the strength of the breakage (Zhang et al., 2000).

The escalating need for natural rubber throughout World War II, coupled with its restricted supply, prompted the exploration of synthetic methods for the production of general-purpose (commodity) rubber. The process of emulsion polymerization was effectively employed in the production of styrene-butadiene rubber (SBR) with the intended application in tire manufacturing. Table 1.1 presents a recipe that has been standardized for the purpose of producing synthetic rubber (Sadhu & Bhowmick, 2004).

Table 1.1. Recipe for the Preparation of Synthetics Styrene-Butadiene Rubber GR-S (Government Rubber–Styrene) by Emulsion Polymerization at 50°C.

Component	Parts By Mass
Styrene	25
Butadiene	75
Lauryl mercaptan	0.5
Water	180
Fatty-acid soap	5
Potassium persulfate	0.3

Source: C.D. Anderson, Creative Commons License.

Diene-based rubbers can undergo vulcanization in a similar manner to natural rubber due to the presence of unsaturated double bonds. Synthetic rubbers are frequently reinforced with inorganic filler materials to augment their mechanical characteristics. Certain diene-based rubbers, like polychloroprene as well as acrylonitrile-butadiene, are designed with the inclusion of supplementary chemical elements within the polymer chain, namely chlorine and nitrogen, respectively, in order to enhance their chemical resistance properties (Mousa, & Karger - Kocsis, 2001).

1.6.2.2. Polyacrylics

The copolymer properties of acrylic monomers can be effectively controlled and customized to meet diverse application requirements owing to their inherent copolymerization ease. Acrylic latexes find frequent application in the coverings, clothing, as well as adhesives sectors. The incorporation of functional groups into acrylic polymers is a facile process, which results in an augmentation of the adaptability of the ultimate product. The copolymer composition can be manipulated to alter both the solvent resistance and glass transition temperature. Adhesives typically exhibit a lower glass transition temperature (Tg) and a softer mechanical behavior, while coatings generally possess a higher Tg as well as greater toughness. The copolymer's tackiness, which refers to its adhesive properties, is closely linked to its composition. The morphology of particles can be altered by means of recipe optimization and the process of polymerization. A variety of surfactants and polymerization techniques can be employed based on the intended outcome.

Acrylic latexes exhibit a high degree of flexibility with regards to both their formulation and application. Table 1.2 presents an illustration of the manufacturing for an acrylic dispersion (Arkaban et al., 2022).

Table 1.2. Recipe for the Preparation of Acrylic Copolymer Dispersion by Emulsion Polymerization at 70° C.

Component	**Parts by mass**
2-Ethylhexylacrylate	50
n-Butylacrylate	50
Water	100
Polyoxyethylenated alkyl-phenol surfactant	3
Ammonium persulfate	0.2

Source: C. D. Anderson, Creative Commons License.

A diverse range of acrylic monomers is at disposal, each exhibiting distinct characteristics. Styrene is frequently employed in the role of comonomer in acrylic latex formulations due to its extensive availability and compatibility. Minor amounts of auxiliary monomers may be employed to confer distinctive characteristics to the latex. In spite of the similarity in reactivity between acrylic monomers and styrene, the occurrence of copolymerization is less probable if there are significant differences in the solubilities of the monomers. This may lead to the development of structured particle morphologies. Semi-continuous polymerization is a viable method for enhancing the uniformity of copolymer composition. In certain instances, there may be a preference for a methodical morphology, which can be intentionally incorporated into the procedure (Wall & Gill, 1954).

Acrylic latexes, which serve as binders, are frequently accompanied by pigments and fillers, although this is not always the case. The system's optical and mechanical properties can be altered based on the quantity and type of coloring or filler material employed, including opacity, gloss, strength, and modulus (Kreuter, 1983).

1.6.2.3. Polyvinyl Acetate

The application of vinyl acetate (VAc) copolymers is primarily observed in the capacity of binders for both indoor and outdoor architectural coatings, as well as adhesive applications. Vinyl acetate monomer is a cost-effective material that exhibits favorable wear resistance properties in its polymerized form, rendering it highly suitable for the aforementioned applications. The efficacy of conventional surfactants in stabilizing dispersions of vinyl acetate is limited. Water-soluble polymers, namely polyvinyl alcohol (PVA) as well as hydroxyl ethyl cellulose (HEC), are commonly employed as protective colloids in order to stabilize vinyl acetate polymerizations. These stabilizers function via steric stabilization mechanism by creating a hydrated boundary layer surrounding the particles undergoing polymerization (Kolter et al., 2013).

Vinyl acetate monomer exhibits high reactivity and notable solubility in water. Due to certain factors, the incorporation of comonomers in copolymerizations with other monomers is frequently inadequate. Under such circumstances, it is common for particles to exhibit structured morphologies, such as core-shell configurations. The utilization of semi-continuous addition has the potential to enhance the copolymer composition. Vinyl acetate latexes can be enhanced in terms of mechanical properties and glass transition temperature by the addition of comonomers such as n-butyl acrylate or methyl methacrylate, belonging to the acrylic and methacrylic families, respectively. In contrast to the polymers with higher hydrophobicity, vinyl acetate latexes can be plasticized by water, which facilitates the process of film formation. However, the incorporation of hydrophobic monomers can be considered to decrease the water sensitivity for the resulting latex films. The copolymerization of ethylene and vinyl acetate can yield ethylene-vinyl acetate (EVA) dispersions that are utilized in various applications such as nonwoven binder, adhesive, and latex paint. Due to the significant exothermic nature of vinyl acetate polymerization, it is recommended to incorporate the monomer in a semi-continuous manner for safety purposes (Koral et al., 1958). The significant dissolution of vinyl acetate with water leads to the formation of numerous oligoradicals during the nucleation phase. These oligoradicals then undergo micellization and aggregation, ultimately resulting in the formation of polymer particles. The occurrence of gel formation is a common outcome in vinyl acetate polymerizations due to grafting reactions facilitated by chain transfer. The presence of residual monomer poses a significant challenge in polyvinyl acetate latexes due to the considerable partitioning of monomer in the aqueous phase. Thus, it is imperative to carefully select the initiator system in order to generate latexes containing little residual monomer, as stated in reference (Knauss & Kenner, 1980).

1.6.2.4. Polyvinyl Chloride

The manufacturing of polyvinyl chloride (PVC) is frequently accomplished through the processes of suspension or emulsion polymerization. The utilization of this material is widespread in the manufacturing of PVC pipes intended for construction purposes. The addition of impact modifiers is a common practice to enhance the strength of the material. PVC finds application in plastisols to dipped coatings, as well as in shower curtains and vinyl leather. In certain applications where the material's flexibility is required to be retained, the use of non-volatile plasticizers is imperative. Due to its high level of hydrophobicity and excellent barrier properties, PVC is a highly effective barrier against oxygen and water vapor in packaging films (Winkler, 1959).

The polymerization of gaseous vinyl chloride monomer is carried out under conditions of high pressure. The insolubility of polyvinyl chloride polymer in its own monomer results in non-classical emulsion polymerization kinetics. The process of polymerization is characterized by significant chain transfer to monomer, and the development of molecular weight is primarily influenced by the reaction temperature as opposed to the concentration of the initiator. As a result, achieving higher molecular weights requires lower reaction temperatures (Asadinezhad et al., 2012).

1.6.2.5. Acrylonitrile

Acrylonitrile, a monomer with high water solubility, is infrequently utilized in emulsion polymerization in isolation. Instead, it is copolymerized with various other monomers to enhance chemical resistance. Acrylonitrile and butadiene are subjected to copolymerization to yield acrylonitrile-butadiene rubber (NBR and nitrile rubber), which serves as a replacement for natural rubber in automotive as well as latex glove applications, particularly in instances where chemical resistance is a crucial factor. Emulsion polymerization is typically carried out at lower temperatures ranging from 5 to 50°C. The process is terminated prior to achieving high conversion in order to minimize the formation of gel (Karp et al., 2017).

Acrylonitrile is frequently present in impact modifiers, including the acrylonitrile-butadiene-styrene (ABS) variety, which is manufactured through emulsion polymerization. The process of seeded emulsion polymerization involves the grafting of polybutadiene seed latex fragments onto styrene and acrylonitrile. The process of shell formation in styrene-acrylonitrile copolymer leads to the spontaneous separation of polybutadiene domains within. The impact modifier particles that ensue are compounded via polystyrene to generate high impact polystyrene (HIPS). The optimization of the impact modification characteristics of latex particles can be achieved by manipulating the butadiene content, particle shape and size, as well as shell molecular weight. Table 6 provides a fundamental composition for an impact modifier utilized in ABS materials (Yang et al., 2004).

1.7. LATEX PARTICLE MORPHOLOGY

The morphology of latex particles is a critical factor in determining the characteristics and efficacy of latex-based materials. The physical characteristics of latex particles, including their size distribution, shape, as well as surface chemistry, have a significant impact on a range of factors like stability, film formation, as well as interaction with other constituents in a formulation (Sundberg & Durant, 2003).

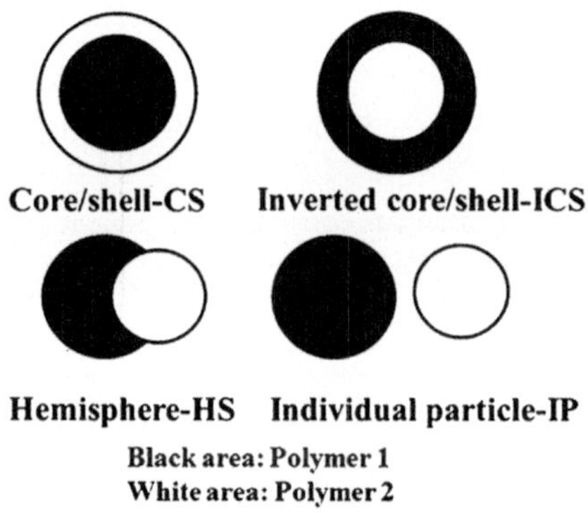

Figure 1.11. Various morphologies for composite latex particles.

Source: Long Bai, Creative Commons License.

1.7.1. Particle Size Distribution

The sizing arrangement of latex fragments is a crucial attribute that has a substantial influence on the properties and conduct of latex-based materials. The term pertains to the spectrum of particle dimensions that are observable in a latex distribution. Various factors such as the process of polymerization, composition of monomers, selection of surfactants, and conditions of reaction can have an impact on the distribution of particle size. In numerous applications, small and monodisperse distributions of particle sizes are preferred due to their ability to improve film uniformity as well as mechanical properties. Conversely, wider size distributions have the potential to cause fluctuations in both the thickness and characteristics of the film. The distribution of particle sizes in latex materials is measured and analyzed through the use of various techniques, including dynamic scattering of light, electron microscopy, as well as laser diffraction (Eliçabe & García-Rubio, 1989).

Remember

Natural latexes are obtained from plants, such as rubber trees, while synthetic latexes are chemically synthesized.

1.7.2. Particle Shape and Morphology

The configuration and structure of latex particles are fundamental variables that impact the physical as well as mechanical characteristics of latex films as well as coatings. Latex particles exhibit various shapes such as spherical, ellipsoidal, rod-like, as well as irregular structures. The morphology of the structure is impacted by various factors, including the conditions of polymerization, interactions with surfactants, and the composition of the monomer. Spheroidal particles are frequently favored owing to their homogeneous dispersion, effective compaction, and favorable film-forming characteristics. Non-spherical particles can offer distinctive functionalities, such as enhanced mechanical reinforcement or improved barrier properties in composite materials. Latex particle shape and morphology are analyzed and characterized using a range of imaging techniques, including scanning electron microscopy (SEM) as well as atomic force microscopy (AFM) (Cho & Lee, 1985).

1.7.3. Surface Chemistry of Latex Particles

The stability, relationship to other components, as well as overall performance of latex particles in various applications are significantly influenced by their surface chemistry. Surfactant molecules or stabilizing agents are commonly found on the surface of latex particles, serving to ensure colloidal stability and hinder aggregation of particles or coagulation. The presence of a surfactant layer on the surface of particles has a significant impact on the electrostatic as well as steric forces that exist between them, ultimately influencing the stability of their dispersion. The efficacy of particles of latex with different additives and substrates is determined by the surface chemistry. As an illustration, the presence of functional groups on the surface of latex particles may enable the occurrence of chemical bonding or attachment with particular surfaces. Surface modification methodologies, like grafting or cross-linking reactions, may be utilized to modify the surface chemistry of particles of latex and incorporate distinct functionalities or augment compatibility with diverse systems (Karaman et al., 1993).

To conclude, it can be stated that the morphology of latex particles, encompassing factors such as size distribution, shape, as well as surface chemistry, exerts a substantial impact on the characteristics and efficacy of latex-based materials. Comprehending and regulating these parameters hold paramount importance in customizing latex formulations to fulfill particular application prerequisites, such as consistency of the film, stability, and bonding. The progress in characterization methodologies and surface modification techniques is consistently contributing towards the advancement of latex materials, which possess improved functionalities and wider scope of applications (Perez-Amodio et al., 2001).

1.8. LATEX FORMULATION AND COMPOSITION

The performance and properties of latex materials in diverse applications are significantly influenced by their formulation and composition. The development of a proficient latex formulation entails the careful selection of polymer types, the integration of surfactants and emulsifiers, as well as the inclusion of coalescing agents as well as film-forming additives to attain the desired properties (Sperry et al., 1994).

1.8.1. Polymer Types in Latex Formulations

A diverse selection of polymers, encompassing both synthetic and natural alternatives, can be utilized in the formulation of latex materials. Natural latexes, such as the one obtained from the fluid of rubber trees, exhibit remarkable properties of flexibility, elongation, as well as tear resistance. These materials are frequently employed in various products such as gloves, balloons, and foam-based items. In contrast, synthetic latexes are generated via the polymerization of emulsions of monomers such as styrene-butadiene, acrylates, and vinyl acetate. Synthetic latexes offer a wider spectrum of characteristics and can be customized to suit particular uses, like coatings, adhesives, as well as textiles. The selection of the polymer category is contingent upon the intended characteristics of the ultimate product. Styrene-butadiene latexes are extensively utilized in coating applications owing to their commendable adhesion, longevity, and hydrophobicity. Acrylic latexes exhibit remarkable weatherability as well as UV resistance, rendering them highly appropriate for employment in outdoor settings. Latex materials can be tailored to meet particular demands for performance by carefully choosing the appropriate polymer type (Steward et al., 2000).

1.8.2. Role of Surfactants and Emulsifiers

Surfactants and emulsifiers play a crucial role in latex formulations by providing stabilization to the distribution of polymer particles in aqueous media. The utilization of surface-active agents aids in the reduction of interfacial tension existing among the polymer as well as water phases, thereby facilitating the formation and sustenance of colloidal latex particles. The utilization of surfactants and emulsifiers in latex formulations yields numerous advantages. They assist in the prevention of particle aggregation as well as coagulation, thereby ensuring stability over an extended period. Latex particles are affected by their environment, which can alter the size of their particles distribution, shape, as well as stability in colloidal form. The optimization of the performance and stability of latex materials may be achieved through the selection of suitable surfactants or emulsifiers (Hellgren et al., 1999).

1.8.3. Coalescing Agents and Film-Forming Additives

The incorporation of coalescing agents as well as film-forming additives into latex formulations is a common practice aimed at facilitating the film creation process and improving the characteristics of the ultimate film. Coalescing agents function as solvents that aid in the amalgamation of latex particles throughout the desiccation and solidification stages, thereby stimulating the creation of a seamless and consistent coating. These substances facilitate the dehydration of the latex film, promoting the proximity of polymer chains and the consequent establishment of a cohesive architecture. Film-modifying agents, including crosslinking agents, plasticizers, and thickeners, are employed to alter the characteristics of the latex film. The incorporation of plasticizers into a film matrix results in an improvement of its mechanical properties, specifically in terms of flexibility and elongation. Crosslinking agents facilitate the establishment of covalent bonds among polymer chains, leading to the augmentation of film robustness, longevity, and immunity to solvents and chemicals. Thickeners are utilized to regulate the viscosity and enhance the application characteristics of the film.

Through meticulous selection of coalescing agents as well as film-forming additives, it is possible to customize latex formulations to attain specific film characteristics, including but not limited to adhesion, flexibility, hardness, as well as chemical resistance (Barbosa et al., 2014). Ultimately, the formulation and composition of latex materials are pivotal factors that dictate their performance and properties. The optimization of stability, film development, and other required characteristics of latex materials in various applications can be achieved through the careful selection of polymer kinds, surfactants, coalescing agents, film-forming additives, and emulsifiers. Through comprehending the function of individual constituents and customizing the composition accordingly, latex substances can be designed to fulfill precise performance criteria (Qin et al., 2015).

1.9. LATEX STABILITY AND STABILITY TESTING

The term "stability" in relation to latex materials pertains to their capacity to sustain a consistent distribution of polymer particles in aqueous medium throughout a given duration. The attainment and sustenance of stability is of paramount importance for the efficacy and durability of products that are based on latex.

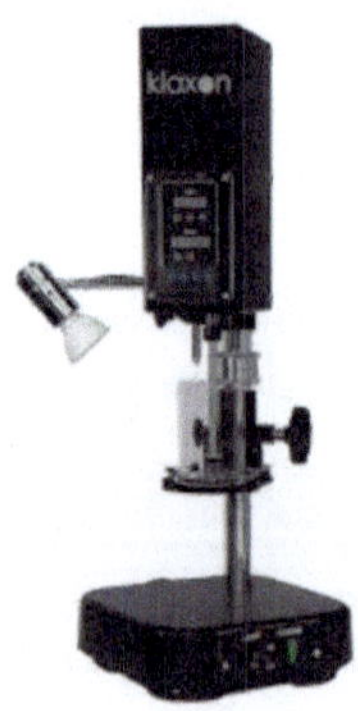

Figure 1.12. Latex mechanical stability testing apparatus.

Source: Latexmst, Creative Commons License.

Did You Know?

atex instability and coagulation can occur due to mechanisms like flocculation and Ostwald ripening.

The durability of latex can be influenced by multiple factors, and stability assessment techniques are utilized to evaluate and track the durability of latex substances. Comprehending the mechanisms underlying latex instability as well as coagulation is a crucial aspect in the development of stable latex-based products (Madge et al., 1952).

1.9.1. Factors Affecting Latex Stability

Following are the factors that affect:

1. The electrostatic charge of particles is a crucial factor in preventing their aggregation, as the repulsion among latex particles is primarily electrostatic in nature. The electrostatic properties of latex particles may be modulated by factors such as pH, ion strength, and the inclusion of electrolytes and additives. Elevated levels of particle concentration may result in heightened probabilities of particle collisions as well as aggregation. The maintenance of stable latex dispersions is contingent upon the effective management of particle concentration (Ortega-Vinuesa & Bastos-González, 2001).
2. The efficacy of surfactants and emulsifiers in maintaining latex particles is of paramount importance for ensuring stability. Inadequate or inefficient coverage of surfactant may lead to the agglomeration of particles and instability (Jacob et al., 2018).
3. Variations in pH as well as ionic strength have the potential to modify the outer charge of latex particulates, thereby exerting an influence on their stability. It is imperative to uphold ideal conditions of pH as well as ionic strength in order to guarantee stability (Paton et al., 1953).
4. Temperature: The stability of latex can be influenced by temperature due to its impact on the rate of aggregate formation or coalescence. Elevated temperatures have the potential to expedite the destabilization mechanisms, ultimately resulting in the instability of latex (Davalos-Pantoja et al., 2001).

1.9.2. Stability Testing Methods

The testing methods are as follows:

1. **Visual Observation:** One method for evaluating the stability of latex dispersions is through visual observation, which involves

inspecting the dispersions over a period of time. This approach is straightforward and efficient. The manifestation of instability can be inferred from alterations in color, appearance, or the emergence of sediment or coagulum (Madge et al., 1951).

2. **Analysis of Particle Size:** Alterations in the distribution of particle size may serve as an indicator of either aggregation of particles or coagulation. Particle size changes can be measured and monitored through the utilization of techniques like dynamic light scattering and laser diffraction (Ren et al., 2019).
3. **Measurement of Turbidity:** Turbidity refers to the degree of opaqueness or transparency exhibited by a latex dispersion. Variations in turbidity levels may suggest the occurrence of aggregation of particles or coagulation. Turbidity measurements can be conducted through the utilization of either turbidimetry or nephelometry techniques (Slotved et al., 2004).
4. **Measurement of Zeta Potential:** The Zeta potential of latex particles is indicative of their surface charge. An enhanced electrostatic repulsion among particles can be attributed to a greater absolute zeta potential, thereby leading to improved stability (Wongkhat & Pattamaprom, 2022).

1.9.3. Mechanisms of Latex Instability and Coagulation

Mechanisms are as follows:

1. **Flocculation:** Flocculation is a phenomenon that arises from the interaction of particles, resulting in the formation of aggregates because of weak attractive forces. Alterations in pH, ionic concentration, or the existence of flocculating agents may give rise to it (Galembeck et al., 2000).
2. **Coagulation:** The process of coagulation entails the permanent amalgamation of latex particles, leading to the creation of a firm conglomeration or coagulum. This phenomenon can be initiated by various factors, including but not limited to temperature variations, alterations in pH levels, or the existence of destabilizing agents (Serra et al., 1992).
3. **Ostwald Ripening:** Ostwald ripening is a phenomenon characterized by the preferential growth of larger particles over smaller ones, which can be attributed to variations in solubility and diffusion rates. The aforementioned mechanism has the

potential to induce alterations in the distribution of particle sizes and consequent instability as time progresses.

Comprehending these mechanisms holds significance in the formulation of stable latex materials and devising approaches to avert or alleviate instability and coagulation (Zhao et al., 2006).

1.10. SCIENTIFIC PRINCIPLES

1.10.1. Colloidal Stabilization

Latex materials are classified as colloids, which implies that they are composed of minute, distinct particles that are dispersed within a continuous liquid medium. Colloidal systems exhibit a plethora of distinctive and captivating characteristics owing to their diminutive dimensions, typically below 1 mm, and substantial interfacial area, typically exceeding 104 cm2/cm3 latex. The significance of surface science in comprehending the formation and stabilization of latex materials is crucial due to the prevailing interfacial effects (Einarson & Berg, 1993).

Tip

Understanding colloidal stabilization, diffusional degradation, and polymerization mechanisms is crucial in studying latex materials.

The process of emulsification involves the dispersal of monomer drops in water to form monomer (M) emulsions, or the formation of latexes through emulsion polymerization. This results in a rise in free energy at the boundary among the colloidal particles as well as the continuous stage. The augmentation of interfacial free energy is significant in colloidal systems characterized by diminutive dimensions and extensive interfacial regions. The thermodynamic favorability of minimizing interfacial free energy serves as a potent driving force for coagulation. The reduction of interfacial area is achieved by droplets through the process of coalescence, leading to the formation of pools of monomer. On the other hand, particles accomplish this by means of aggregation, resulting in the formation of a lot of polymers (coagulum). Evidently, the aforementioned outcomes are not desirable, thus it is imperative to establish a method of ensuring colloidal stability to prevent coagulation. The utilization of a surfactant, also referred to as a stabilizer or emulsifier, serves the aforementioned objective (Yokoyama et al., 1988).The molecules of surfactants exhibit amphiphilic properties, indicating the presence of both hydrophilic (water-soluble) and hydrophobic (water-insoluble) segments. This characteristic has been documented in literature. The partitioning of surfactant molecules among the water as well as monomer/polymer stages is attributed to their amphiphilic nature, which leads to their adsorption at the droplet/particle junctions. Upon adsorption at the interface, the surfactant induces electrostatic and/or steric stabilization mechanisms that result in the separation of droplets or particles by repelling them from each other. It has been observed that a surfactant can serve the dual purpose of functioning as both an electrostatic as well as a steric stabilizer, thereby being classified as an electrosteric stabilizer (Verhaar, 1959).

The electrostatic stabilization process involves the dissociation of surfactant molecules in a solution, leading to the provision of an ionic charge upon adsorption at the surface of a droplet. The particles that are charged on the surface exhibit electrostatic repulsion towards each other. Anionic surfactants bearing a negative ion charge, are juxtaposed with cationic surfactants bearing a positive ionic charge. The incorporation of electrolyte to a latex may cause coagulation due to significant alterations in its ionic strength. Similarly, coagulation can also be induced by the combination of surfactants with opposing charges (Khan & Armes, 2000).

The steric stability mechanism involves the utilization of surfactant molecules that are polymeric chains with water solubility and a degree of conformational mobility. Upon the approach of two particles coated with nonionic surfactant, the adsorbed layer experiences compression, leading to a reduction in the ability to move of the stabilizer chains. The particles exhibit a thermodynamically unfavorable rise in free energy, leading to their mutual repulsion.

Surfactants exhibit surface activity, thereby reducing the interfacial tension among the water as well as monomer phases. The reduction in interfacial tension facilitates the formation of smaller droplets during the dispersion process of the monomer in the aqueous phase. Furthermore, the thermodynamic impetus for the process of coalescence is diminished due to the decrease in interfacial tension. Emulsions that are prepared with surfactants exhibit greater colloidal stability, primarily due to the electrostatic as well as steric stabilization mechanisms (Rogach et al., 2000).

1.10.2. Diffusional Degradation

In addition to coalescence, emulsions can undergo diffusional degradation, a mechanism that results in the reduction of the number of droplets and an increase in their size. The method of interfacial free energy minimization causes the monomer to diffuse from smaller droplets to larger ones. The aforementioned occurrence is commonly referred to as Ostwald ripening, as documented in literature.

The presence of emulsion droplets is crucial in facilitating effective mass transfer during the process of emulsion polymerization, owing to the significant interfacial area they offer. The majority of monomers exhibit a degree of solubility in water, facilitating their transfer throughout the aqueous state from the monomer drops to the polymer particles, where polymerization takes place. In the process of polymerization, the concentration gradient of monomers will surpass the Ostwald ripening forces, leading to the diffusion of monomers from larger drops to smaller monomer-swollen particles (Chern et al., 1998).

The inclusion of a costabilizer can mitigate the degree of monomer diffusion. In order for Costabilizer to exhibit optimal droplet diffusional stability, it is imperative that it possess two distinct characteristics: a low molecular weight to enhance its combined free energy contribution, and low water solubility to impede its transport throughout the aqueous phase. The process of incorporating costabilizer into monomer droplets prior to dispersion is employed to counteract the phenomenon of Ostwald ripening that may occur prior to polymerization. This technique is predominantly utilized in the preparation of miniemulsions (Jasinski et al., 2014).

1.10.3. Free Radical Polymerization

The majority of emulsion polymerizations are classified as free radical mechanisms. The free radical polymerization mechanism comprises of a series of distinct stages, namely initiation, propagation, as well as termination. During the initial stage, a chemical compound known as an initiator undergoes thermal decomposition to produce free radicals. The rate of initiator decomposition can be characterized via an Arrhenius-type equation which incorporates a decomposition constant (kd) which is the inverse of the initiator's half-life (t1/2). Free radicals trigger the process of polymerization through their reaction with a nearby molecule of monomer. The occurrence marks the initiation of a fresh polymer chain. Due to the continuous decomposition of initiator molecules into radicals, there is a perpetual generation of fresh polymer chains. The monomeric molecules that have undergone initiation possess an active end group that is free radical in nature. In the process of propagation, the monomeric species that have been initiated encounter uninitiated monomer molecules as well as undergo a reaction that results in the formation of dimers that possess active end groups. The process of oligomerization is initiated by the reaction of dimers with monomers, as reported in previous studies. The process of oligomeric chain growth occurs through propagation, leading to a subsequent increase in molecular weight. The relationship between the rate of growth and the propagation rate constant (kp) is directly proportional. The propagation rate constant (kp) varies for each monomer and is subject to an Arrhenius dependence on temperature. The polymerization rate (Rp) is equivalent to the rate at which growth occurs of the polymer chains.

Termination is the cessation of polymer chain growth that occurs upon deactivation of the free radical end group. The process of termination in polymerization can occur through two mechanisms: combination, wherein both active radical end groups come into contact, or disproportionation, wherein the active radical is eliminated from a developing polymer chain through the removal of hydrogen from a different growing chain. Chain growth can be terminated through chain transfer to another species, such as a monomeric or polymeric entity.

The occurrence of branching and crosslinking reactions, leading to gel formation, can be attributed to intermolecular transfer of chains to the polymer, as stated in reference 343. The termination rate is directly proportional to the removal rate constant (kt) that exhibits an Arrhenius dependence on temperature. At elevated levels of monomer-to-polymer conversion, the inner viscosity of the particles of latex becomes significant. Under these circumstances, there is a significant decrease in the termination rate, while the polymerization rate experiences an acceleration.

The phenomenon of autoacceleration is commonly referred to as the Tromsdorff gel impact in academic circles. While free radical polymerization is the most prevalent method, alternative polymerization techniques have been implemented in emulsion polymerization. These include reversible addition-fragmentation transfer (RAFT), atom transfer radical polymerization (ATRP), and stable free radical polymerization (SFRP) (Noble & Coote, 2013).

1.10.4. Particle Nucleation Mechanisms

The utilization of surfactant is known to effectively maintain the colloidal stability of emulsion droplets as well as latex particles, thereby preventing coalescence and aggregation. In addition to its stabilizing function, the surfactant serves a crucial role in emulsion polymerization. The involvement of surfactant in the nucleation mechanism of polymer latex particles is of critical importance. The precise quantity of surfactant employed plays a pivotal role in regulating the size distribution of latex particles. Upon addition of surfactant to an emulsion, a portion of it dissolves in the aqueous phase while the remaining fraction adheres to the surface of the emulsion droplets, conforming to an adsorption isotherm such as Freundlich, Langmuir, or Frumkin adsorption isotherms.

Upon the escalation of the free surfactant amount in the aqueous phase, a saturation threshold is eventually attained, beyond which any further surfactant becomes insoluble. The aforementioned juncture is commonly referred to as the critical micelle concentration (CMC). Upon reaching the CMC, any surfactant introduced will undergo association to form micelles, as documented in literature.

The hydrophobic cores of the micelles exhibit an affinity for the monomer present in the stabilized droplets, leading to the expansion of the micelles. Oligoradicals are formed through the reaction of radicals produced by the initiator with monomer that is dissolved in the aqueous phase. Upon reaching a critical chain length, oligoradicals have the potential to undergo various nucleation processes to form primary particles.

These processes include homogeneous nucleation, micellar nucleation within monomer-swollen micelles, and droplet nucleation within monomer droplets. It is worth noting that the rate of droplet nucleation is relatively low due to the limited number of droplets present. Figure 1.13 provides a visual representation of the diverse potential mechanisms for particle nucleation. Additional theories regarding nucleation mechanisms in a solution polymerization, like coagulative nucleation, have been proposed to further elaborate upon these foundational principles.

Certain emulsion polymerizations are conducted in the absence of surfactants. The aforementioned polymerizations are frequently denoted as emulsifier-free polymerizations.

Persulfate initiators are frequently employed in polymerizations to initiate the process via uniform nucleation in the aqueous phase. The process of oligoradical aggregation leads to the formation of particles that are stabilized through electrostatic colloidal means, facilitated via the persulfate initiator end groups that arise from surface initiation. Surface-active monomers, specifically sodium styrene sulfonate, are utilized in limited amounts to provide supplementary electrostatic stabilization (Weber et al., 1996).

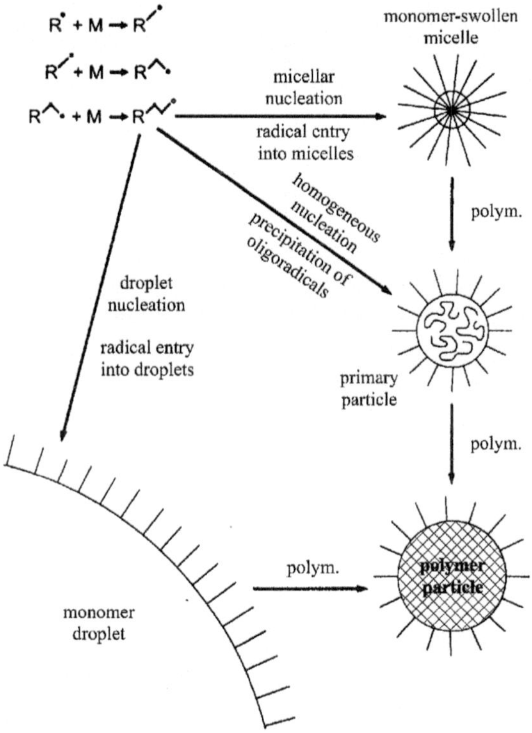

Figure 1.13. Homogeneous, micellar, and droplet nucleation mechanisms.

Source: C.D. Anderson, Creative Commons License.

1.10.5. Emulsion Polymerization Intervals

According to Harkins' model of emulsion polymerization, the process of polymerization can be theoretically broken down into three intervals (as illustrated in Figure 1.14), wherein polymer particles are nucleated as well as subsequently increase in size. In Interval I, the process of particle nucleation occurs through either radical entry into micelles or homogeneous nucleation. As the formation of new particles takes place, there is a simultaneous increase in both the quantity of particles as well as the rate of polymerization. The utilization of surfactant in stabilizing nascent particles results in a decrease of the concentration of unbound surfactant below the threshold level of micelle formation, thereby terminating the process of nucleation. Subsequent to this juncture, the quantity of particles remains invariant.

In Interval II, the extant particles undergo further polymerization and utilize the monomer present in the sizable monomer droplets. The transportation of the monomer to the polymerization site, where the polymer particles are growing, occurs via the aqueous phase due to the presence of a concentration gradient. Upon the conclusion of Interval II, the monomeric units contained within the droplets have been exhausted, resulting in the absence of any further droplets containing monomers.

In Interval III, the process of polymerization continues within the polymer particles until the monomer is completely transformed into polymer. The monomer concentration within the polymer particles, denoted as [M]p, diminishes to a negligible level.

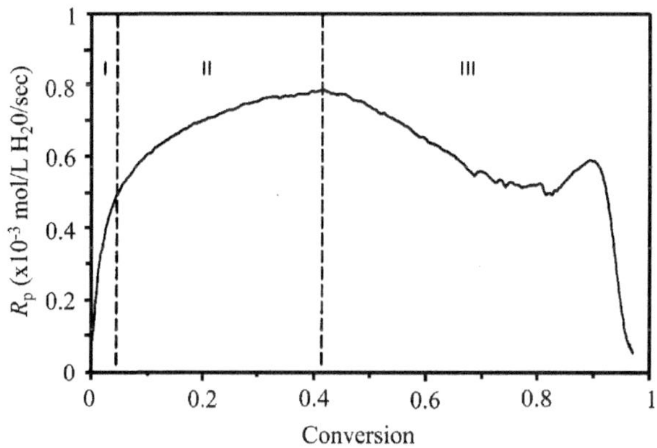

Figure 1.14. Intervals in emulsion polymerization.

Source: C. D. Anderson, Creative Commons License.

It is possible that a Tromsdorff effect occurring towards the finish of the polymerization process could lead to an autoacceleration in the speed of polymerization (Ugelstad & Hansen, 1976).

1.10.6. Smith-Ewart Kinetics

The speed at which a polymerization takes place is referred to as the polymerization's kinetics. The primary polymerization kinetics theory for emulsion polymerization is that of Smith as well as Ewart. They put forth a theoretical model in which radicals continuously enter polymer particles that have been swollen by monomers (radical flux). The radicals may initiate polymerization within the particles, eliminate from the particle, or terminate with one another. Immediately coming to an end, and n might be much larger than 1. In essence, these polymerizations adhere to bulk polymerization kinetics.

The first scenario (Case I), in which radical desorption from the particulate is probable and the particles are tiny or the monomer is significantly water-soluble, has a very low n and a slow polymerization rate. The likelihood of a radical exit is minimal in Case II, the second scenario. Zero-one kinetics states that once a radical enters a tiny particle, polymerization continues till a second radical arrives, at which point both are instantly terminated. N is equivalent to 1/2 in this scenario. The particles in the third scenario, Case III, are big enough to allow multiple radicals to coexist in it without

Smith and Ewart used a surfactant balance to find that the speed of polymerization is inversely proportional to the initiator amount to the 0.4 power and the amount of surfactant to the 0.6 power. In his experiments illustrating the idea, Smith used sodium dodecyl sulfate (SDS) to be the surfactant and styrene as the monomer. The majority of monomer/surfactant systems don't follow the ideal Smith-Ewart kinetics in real life. But across all emulsion polymer systems, equation below holds true:

$$R_p = \frac{k_p [M]_p N_p \bar{n}}{N_A}$$

For three different scenarios where the particle size as well as radical exit rate may differ, the mean quantity of radicals per particle (n) was determined through mathematical analysis (Hawkett et al., 1977).

1.11. LATEX FILM FORMATION MECHANISMS

While a latex dispersion is used as well as dried, creating a continuous as well as cohesive film, this crucial process of latex film formation takes place. To achieve desired film features in a variety of applications, it is crucial to comprehend the processes of latex formation, the significance of temperature as well as drying conditions, and also the resulting film's structure and properties (Dobler & Holl, 1996).

1.11.1. Latex Coalescence and Film Formation

Several processes, including the coalescence of particles of latex and the resulting creation of a continuous film, take place during the formation of latex films.

1. **Fusion of Latex Particles:** The evaporation of water and other volatile solvents causes a rise in particle concentration as the latex dispersion is put on to a substrate. The contact as well as fusion of nearby latex particles is made easier by the higher particle density.
2. **Rupture of Boundaries and Bridging:** Interfacial forces like Van der Waals attraction as well as capillary forces can cause temporary bridges to form if latex particles arrive into contact. These bridges eventually break down, permitting the particles to combine and create a continuous film (Winnik, 1997).
3. **Redistribution of Polymer Chains and Coalescence:** Polymer chains rearrange as well as redistribute across the film as latex particles fuse, creating intermolecular connections. This procedure encourages the development of a strong structure and strengthens the integrity of the film (Chevalier et al., 1992).

1.11.2. Role of Temperature and Drying Conditions

The kinetics as well as mechanisms involved in the formation of latex films are significantly influenced by temperature and drying conditions:

1. **Evaporation Rate:** Water or solvents in the latex film evaporate faster at greater temperatures as well as lower humidity levels. Faster latex particulate coalescence as well as film formation may result from rapid evaporation.
2. **Film Thickness and Drying Time:** The film thickness as well as properties are impacted by the drying time. Thicker films are produced as a result of more broad coalescence and chain rearrangement during slower drying. On the other hand, quick drying can produce thinner films with unique properties.
3. **Effects of Temperature on Polymer Mobility:** Polymer chains become more mobile at higher temperatures, which facilitates their diffusion as well as rearrangement during the formation of films. This improved mobility aids in the formation of a film structure that is more uniform and strong (Routh & Russel, 1999).

1.11.3. Latex Film Structure and Properties

The latex composition, size of particles, drying conditions, as well as additives are just a few of the variables that have an impact on the film structure as well as properties of latex films. Important characteristics are:

1. **Mechanical Strength:** Latex films typically have strong mechanical properties because of the cohesive structure that coalescence creates. Crosslinking agents and additives can be used to further increase film strength.
2. **Adhesion:** Due to interfacial interactions as well as the latex particles' capacity to adapt to the substrate surface, latex films frequently display excellent adhesion to a variety of substrates.
3. **Barrier Properties:** Latex films can display a variety of barrier qualities, like resistance to water, gas permeability, as well as durability against chemicals or solvents, based on the composition as well as film thickness.
4. **Surface Morphology:** Depending on elements like latex particle size as well as drying conditions, the surface shape of latex films may vary from smooth to textured. Surface features can be altered by adding substances or applying after-treatments.
5. Knowing the film structure as well as features of latex films makes it possible to formulate materials with specialized functionalities for particular applications, including coatings, adhesives, as well as barrier films.

In conclusion, the assembly of latex particles into a continuous film is what causes latex film formation. The coalescence process as well as the resulting film's structure and properties are influenced by variables like temperature, drying conditions, as well as latex composition. Latex films can display the desired properties, such as rigidity, adhesion, barrier characteristics, and surface morphology, by optimizing these parameters (Zosel & Ley, 1993).

1.12. RHEOLOGY OF LATEX SYSTEMS

Understanding the flow as well as deformation behavior of latex systems, also known as rheology, is crucial for comprehending their processing and usage characteristics. Characterizing and improving latex formulations requires consideration of the viscoelastic properties of latexes, the impact of the concentration of particles and size, and different rheological measurement techniques (Hester & Squire, 1997).

1.12.1. Viscoelastic Behavior of Latexes

Latex materials display viscoelastic characteristics, which involve a combination of properties from both elastic solids and viscous liquids. The observed behavior can be attributed to the existence of polymer chains as well as the inter-particle interactions within the latex system. The fundamental components of viscoelastic behavior in latex systems comprise:

1. **Elasticity (Storage Modulus, G'):** The viscoelastic behavior's elastic component pertains to the material's capacity to retain and restore energy during deformation. This pertains to the viscoelastic properties exhibited by latex materials, which demonstrate their capacity to withstand deformation and exhibit elastic recovery.
2. **Viscosity (Loss Modulus, G"):** The viscoelastic behavior of a material is characterized by its ability to eliminate energy as heat during deformation, which is attributed to its viscous component. This pertains to the fluidic characteristics of latexes, denoting their rheological properties encompassing both their flow and viscoelasticity.
3. **Viscoelastic Spectrum:** The frequency-dependent viscoelastic spectrum is commonly utilized to characterize the viscoelastic properties of latexes. This spectrum illustrates the changes in elastic and viscous components with respect to frequency or time.

Comprehending the viscoelastic properties of latexes is of paramount importance in forecasting their fluidity, coating efficacy, stability, and reaction to extrinsic forces throughout their processing and implementation (Kan & Blackson, 1996).

1.12.2. Influence of Particle Concentration and Size

The rheological characteristics of latex systems are significantly impacted by both the concentration and size of particles. The dispersion characteristics of latex, interparticle interactions, as well as the consequent rheological behavior are influenced by these parameters. Significant observations encompass:

1. **Concentration Effects:** Elevating the particle amount in latex systems typically results in heightened viscosity and elastic characteristics owing to augmented particle-particle interactions and polymer chain entanglement. The flow and coating efficacy of latex formulations may be impacted by this phenomenon.
2. **Particle Size Effects:** The rheology in latexes is directly affected by the particle size. Reduced viscosity and enhanced flowability are commonly observed in smaller particles, whereas larger

particles tend to result in elevated viscosity and greater elasticity. The rheological properties of a substance are also affected by the distribution of particle sizes.

Comprehending the impact of the concentration of particles and size on the rheological behavior of latex systems facilitates the process of formulating optimal compositions, fine-tuning flow characteristics, and devising coatings or products with targeted attributes (Macchini et al., 2013).

1.12.3. Rheological Measurement Techniques

Diverse methodologies are employed to evaluate the rheological properties of latex systems. Several frequently utilized techniques include:

1. **Rotational Viscometry:** The methodology entails the quantification of the rotational force necessary to turn a spindle that is submerged in the latex specimen at a designated rate of shear. This document presents data pertaining to the viscosity as well as flow characteristics of latex materials.
2. **Oscillatory Rheometry:** Oscillatory shear tests are a method of analyzing the viscoelastic properties of a latex sample. This is achieved by subjecting the sample to changing shear strains and stress, which enables the determination of parameters such as the storage modulus (G'), modulus of loss (G"), as well as complicated viscosity.
3. **Dynamic Mechanical Analysis (DMA):** viscoelastic characteristics of latex films or coatings are evaluated through the application of controlled mechanical deformation at varying frequencies and temperatures, which is known as DMA. The aforementioned statement offers valuable perspectives on the mechanical properties and viscoelastic characteristics of the material.

The utilization of measurement techniques facilitates the characterization of flow, deformation, as well as viscoelastic characteristics of latex systems, thereby enabling optimization of formulation and customization of rheological properties to suit specific applications (Collyer, 1993).

1.13. LATEX MODIFICATION AND FUNCTIONALIZATION

The properties of latex materials may be enhanced and their application opportunities can be expanded through modification and functionalization.

The implementation of crosslinking techniques on latex films, surface modification methodologies, and the blending and copolymerization of latex are commonly employed methods to attain targeted modifications and functionalizations.

1.13.1. Crosslinking of Latex Films

Crosslinking is a process that entails the establishment of chemical bonds among polymer chains, leading to the creation of a three-dimensional structure of networks inside the latex film. Crosslinking can be accomplished through diverse techniques, such as:

1. **Chemical Crosslinking:** Chemical crosslinkers, which are inclusive of multifunctional monomers and cross-linking agents, are incorporated into the latex creation. Throughout the process of film formation and desiccation, the crosslinkers undergo a chemical reaction alongside the polymer chains, resulting in the creation of covalent bonds and the establishment of a crosslinked network.
2. **Physical Crosslinking:** Physical crosslinking is a process that is dependent on the non-covalent interactions among polymer chains, which may include crystallization or hydrogen bonding. The aforementioned interactions facilitate the formation of transient or invertible crosslinks, thereby augmenting the robustness and durability of the film.

The mechanical characteristics, durability, solvent resistance, as well as dimensional durability of latex films are improved through the process of crosslinking. Furthermore, it enhances their capacity to withstand variations in temperature, exposure to ultraviolet radiation, and chemical deterioration (Winnik, 2002).

1.13.2. Surface Modification Techniques

Surface modification methodologies are utilized to modify the surface characteristics of latex materials, thereby conferring particular functionalities or augmenting how they interact with other constituents. Frequently employed methods for surface modification encompass:

1. **Chemical Modification:** Chemical transformations, including grafting and functional group alterations, are executed at the interface of particles of latex or films. The aforementioned

reactions facilitate the incorporation of intended functionalities, including but not limited to hydrophobicity, hydrophilicity, adhesion advancement, or targeted chemical reactivity.

2. **Physical Surface Treatments:** Various physical techniques such as plasma therapy, corona treatment, and UV irradiation can be utilized to alter the surface characteristics of latex materials. The aforementioned treatments have the ability to modify the surface energy, imperfections, and topography, thereby resulting in enhanced adhesion, wettability, and barrier properties.

The utilization of surface modification techniques facilitates the modification of latex materials to cater to specific applications. This includes the enhancement of adhesion to substrates, customization of wetting properties, and improvement of compatibility with other elements in a formulation (Min et al., 2002).

1.13.3. Latex Blending and Copolymerization

The process of blending and copolymerization entails the amalgamation of distinct latexes or monomers to generate materials that exhibit distinctive properties and performance attributes. Various methodologies have been employed, encompassing the following strategies:

1. **Latex Blending:** Various latexes possessing complementary properties, like distinct particle sizes and polymer pieces, may be amalgamated. The amalgamation of favorable characteristics from individual latex varieties facilitates the enhancement of efficacy or customization of attributes.
2. **Copolymerization:** During the process of emulsion polymerization, it is possible to copolymerize monomers that possess distinct chemical structures and functional groups. The outcome of this process is the generation of copolymer latexes that exhibit distinct compositions, properties, and reactivities in comparison to those of homopolymer latexes.

The utilization of latex blending as well as copolymerization techniques provides the opportunity to precisely adjust characteristics such as mechanical robustness, pliability, bonding capability, and chemical resilience. They offer prospects for fabricating hybrid latex materials which fulfill particular prerequisites for diverse applications (Zhao & Urban, 2000).

1.14. FUTURE TRENDS AND EMERGING TECHNOLOGIES IN LATEX MATERIALS

The domain of latex materials is subject to constant evolution, as persistent research, and development efforts drive progress in the areas of synthesis, processing methodologies, and the investigation of novel materials. Prospective tendencies and nascent technologies in the domain of latex materials encompass progressions in the synthesis and processing methodologies of latex, the amalgamation of the use of nanotechnology on latex uses, as well as the innovation of bio-based and sustainable latex materials.

Tip

Advances in latex synthesis, nanotechnology applications, and bio-based materials contribute to the future development of latex materials.

1.14.1. Advances in Latex Synthesis and Processing Techniques

Scholars and scientists are perpetually investigating novel methods of synthesis and processing with the aim of enhancing the efficacy, functionality, and eco-friendliness of latex materials. Several areas of concentration encompass:

1. **Controlled Polymerization Techniques:** The development of controlled polymerization techniques, like controlled/living radical polymerization, has facilitated the accurate regulation of polymer construction, molecular weight, as well as functionality. The utilization of these methodologies facilitates the amalgamation of latexes that possess customized characteristics and enhanced regulation over particle dimensions, dispersion, and structure.
2. **Green and Sustainable Synthesis:** There is a current endeavor to establish techniques that are both ecologically sound and sustainable for the production of latex. The aforementioned practices encompass the utilization of sustainable raw materials, biologically derived building blocks, and environmentally conscious methodologies that mitigate the employment of harmful substances and curtail energy expenditure.
3. **Advanced Processing Methods:** The objective of advancements in latex processing methodologies is to enhance the efficacy of

film development, coating application, as well as the expandability of production procedures. Novel methods like electrostatic depositing, 3D printing, as well as advanced coating techniques present promising opportunities for accurate deposition and improved film characteristics (Scott et al., 2020).

1.14.2. Nanotechnology in Latex Applications

The incorporation of nanotechnology into latex-based products presents novel opportunities for improved characteristics and capabilities. Several significant domains of growth and progress encompass:

1. **Nanoparticle Reinforcement:** The integration of nanoparticles, like silica, carbon nanotubes, and graphene, to latex formulations has been found to enhance mechanical strength, barrier characteristics, as well as conductivity. The nanocomposite latex materials exhibit promising potential for utilization in various domains including but not limited to flexible electronic devices, high-performance coatings, as well as advanced textiles.
2. **Stimuli-Responsive Latexes:** Smart and responsive latexes may be developed by integrating stimuli-responsive nanoparticles and materials into latex formulations. These materials are capable of displaying responsive properties to external stimuli such as changes in temperature, pH, and light. The aforementioned entities exhibit potential utility in various domains such as sensor technology, drug administration mechanisms, and materials capable of self-repair (Fernando & Sung, 2009).

1.14.3. Bio-based and Renewable Latex Materials

The increasing focus on sustainability as well as the requirement for substitute raw materials has resulted in the emergence of bio-derived and sustainable latex materials. Significant advancements in this field encompass:

1. **Natural and Bio-based Latexes:** The exploration of natural latexes as well as bio-based polymers obtained from renewable sources, including plant-based sources and biomass, has been intensified. This provides eco-friendly substitutes to synthetic latexes. The aforementioned materials exhibit promising characteristics that could potentially serve as substitutes for latexes derived from petroleum in diverse applications.

2. **Green Processing and Additives:** There is a current endeavor to create environmentally-friendly processing techniques and incorporate bio-derived additives into latex compositions. The implementation of environmentally friendly surfactants, crosslinkers that are capable of biodegradation, and manufacturing processes that prioritize sustainability are all part of the efforts to reduce the negative impact on the environment.

The industry endeavors to promote sustainable practices, minimize reliance on fossil fuels, and decrease carbon footprint by integrating bio-based as well as renewable resources into latex formulations (Samantarai et al., 2019).

1.15. LATEX RECYCLING AND ENVIRONMENTAL CONSIDERATIONS

The promotion of sustainability and waste reduction necessitates the recycling of latex content as well as the mitigation of their environmental impact. The recycling of latex poses certain challenges and necessitates the implementation of strategic measures, whereas the disposal of latex-based items may result in environmental ramifications. Furthermore, the implementation of sustainable techniques in both the production and utilization of latex serves to mitigate the ecological impact of the aforementioned sector (Mohajerani et al., 2020).

1.15.1. Challenges and Strategies for Latex Recycling

Below are the strategies and challenges:

1. **Material Complexity:** The recycling of latex products poses a challenge due to the intricate composition of polymers, additives, as well as fillers present in them. Innovative methods and advances in technology are necessary for the efficient separation and processing of distinct components to facilitate their recycling.
2. **Contamination and Cross-Contamination:** The presence of extraneous substances, like adhesives or impurities originating from products at the end of their useful life, can impede the process of recycling. Efficient techniques for the separation and purification of latex substances are imperative in preserving the caliber of repurposed latex.

3. **Collection and Sorting Infrastructure:** The implementation of effective collection systems as well as sorting infrastructure is imperative for the prosperous recycling of latex products. The development and implementation of recycling programs necessitates the cooperation of producers, waste management companies, as well as regulatory bodies.
4. **Market Demand and Product Design:** The long-term success of recycling initiatives is contingent upon the establishment of a market need for recycled latex products. Promoting product design that enables recyclability, through the utilization of compatible materials and simplification of the product structure, can effectively bolster recycling initiatives (Kheirabadi & Sheikhi, 2022).

1.15.2. Environmental Impacts of Latex Disposal

Below are the environmental effects:

1. **Landfill Accumulation:** Improper disposal of latex-based goods may result in their buildup in landfills, thereby occupying significant space and exacerbating the overall waste burden. Latex-based materials are known to exhibit non-biodegradable properties, thereby leading to their prolonged persistence in landfills.
2. **Leaching of Chemicals:** Certain latex-based items may comprise of additional substances or compounds that have the potential to permeate the surroundings if not disposed of in an appropriate manner. The presence of these chemicals has the potential to cause contamination in soil, water sources, as well as ecosystems, thereby posing significant risks to the health of humans and the environment.
3. **Greenhouse Gas Emissions:** The disposal of latex materials through incineration has the potential to generate greenhouse gas emissions, such as carbon dioxide, as well as air pollutants. The aforementioned emissions are known to be a significant contributor to both the phenomenon of climate change and the issue of air pollution (Wang et al., 2022).

1.15.3. Sustainable Practices in Latex Production and Use

Below are the sustainable practices:

1. **Raw Material Selection:** The selection of renewable as well as bio-based substances for the production of latex diminishes

the dependence on non-renewable energy sources and fosters the principles of sustainability. The aforementioned entails the investigation of substitute origins and the formulation of environmentally sustainable methods for the cultivation and collection of natural latex.

2. **Energy and Resource Efficiency:** The implementation of energy-efficient processes and the optimization of the use of resources in latex production are effective measures to minimize the environmental impact. This entails the utilization of sustainable energy sources, enhancement of operational efficiency, and reduction of waste production.
3. **Life Cycle Assessment (LCA):** Undertaking LACs on latex materials facilitates a thorough appraisal of their ecological repercussions. The utilization of LCA facilitates the identification of potential areas for enhancement and aids in the process of making informed decisions that promote the adoption of more environmentally friendly manufacturing and consumption practices.
4. **Circular Economy Approaches:** Incorporating circular economy principles, including the integration of recyclability in design, the establishment of take-back initiatives, and the advocacy for the utilization of recycled latex materials, is conducive to waste reduction and the optimization of resource consumption.
5. **Environmental Certifications and Standards:** Conforming to established environmental certifications and norms guarantees adherence to sustainable practices and exhibits dedication towards responsible production and utilization of latex.

The latex industry has the potential to make a substantial contribution towards a sustainable and environmentally friendly future by tackling recycling challenges, reducing the environmental impact of disposal, and implementing sustainable practices for manufacturing and usage. Effective implementation of strategies and attainment of favorable environmental outcomes necessitate collaboration among stakeholders, encompassing manufacturers, consumers, as well as regulatory bodies (Rasutis et al., 2015).

1.15.4. Environmental Legislation

Historically, polymers have been applied through solution deposition in solvents that are organic for the purpose of coatings and in diverse industrial procedures. VOCs refer to diminutive organic molecules that undergo vaporization and are subsequently released into the atmosphere. The reduction of VOC concentrations in coatings as well as other products is driven by both environmental and safety and health factors. From an environmental standpoint, it has been determined that certain VOCs have a detrimental impact on the ozone layer, while others may potentially contribute to the presence of greenhouse gases that are associated with the phenomenon of global warming. The utilization of organic solvents in coatings has resulted in the generation of harmful oxidants that pose a threat to the environment, as reported by Wicks et al. in 1992. The utilization of solvent-borne products in occupational settings may result in an unpleasant odor. Furthermore, prolonged contact with solvents has been linked to health concerns among workers, particularly those in the professional painting industry, as evidenced by studies conducted by Bockelmann et al. (2002), Steinhauer et al. (2001), and Morrow and Steinhauer (1995). There is a growing trend among consumers to favor household goods that do not release VOCs.

In response to the issue of pollution, governmental bodies have implemented regulatory measures and initiatives aimed at preventing pollution (DeVito, 1999). The pertinent legislation in Europe is the EU Directive 2004/42/EC, which has been transposed into British law through The Organic Volatile Compounds in Paints, Varnishes as well as Vehicle Refinishing Products Regulations 2005. As per the EU Directive, a VOC is defined as a substance made up of organic matter that possesses a boiling point equal to or less than 250°C at a standard pressure in the atmosphere of 101.3 kPa. The permissible thresholds for VOC concentration are contingent upon the intended use. The legislation will be fully implemented in 2010, notwithstanding the fact that the VOC restrictions were already lowered in 2007. As an illustration, the VOC threshold for water-based matt indoor paint was established at 75 grams per liter in 2007, and is expected to decrease to 30 g each liter by 2010.

The majority of coating producers are endeavoring to diminish their levels of VOCs to levels below the prescribed limits. This is not only to adhere to legal requirements, but also to gain a competitive edge in light of the anticipated stricter regulations and to meet the expectations of consumers. According to Dennington (2007), a product's appeal and value can be enhanced by its low VOC content. Numerous nations have implemented distinctive labeling systems to distinguish products deemed to be environmentally sustainable. According to Jotischky (2000), legislation has been put in place to establish "occupational exposure limits" for hazardous chemicals in various manufacturing sectors, including automotive finishing, to safeguard the well-being of workers. The regulations have an impact on the utilization of chemicals within the occupational setting.

The shift from solvent-cast processing of polymers to waterborne processes has played a significant role in propelling the expansion of the latex market. In the past, it was customary to utilize solvent-based household paints. The cleaning of brushes was carried out by means of mineral spirits and turpentine. Presently, the majority of architectural coatings employ latex compositions, while exclusively certain varnishes are solvent-based. However, it is noteworthy that even formulations of latex may comprise a considerable amount of VOCs. A range of diminutive molecules are utilized to act as plasticizers, thereby inducing softness in the polymer present in the latex particles. This facilitates the formation of a film at ambient temperature or below. Upon the full evaporation of the VOC, the polymer undergoes a transition back to its rigid, un-plasticized form, resulting in the formation of a polymer film with a high glass transition temperature (Tg). The production of hard latex coatings that do not emit VOCs poses a significant challenge, necessitating ongoing research into the formation of latex films. The aforementioned necessity imbues scientific endeavors with a heightened sense of exigency and has, in part, served as a motivating factor for the composition of this literary work.

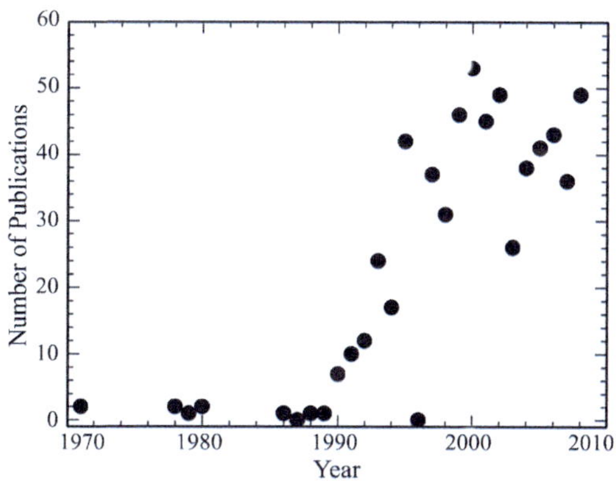

Figure 1.15. Number of publications, by year, on the topic of 'latex film formation' according to a key word search on the ISI Web of Knowledge database.

Source: E. S. Daniels, Creative Commons License.

The scientific literature has witnessed a surge in desire to latex film formation, which could be attributed to the environmental and economic concerns. Figure 1.15 provides a historical viewpoint. Prior to 1990, a limited number of publications were dedicated to the examination of latex film formation. Throughout the 1990s, there had been a consistent upward trend in the annual quantity of publications, culminating in a zenith of fifty-three in the year 2000. During the initial decade of the 21st century, there has been a fluctuation in the figures, which have centered on a mean of 41 annually (Datta & BV, 2004).

Activity 1.1

You are a research scientist tasked with formulating a new latex coating. How would you select appropriate latex materials, ensure stability, and address environmental considerations in the production process?

1.16. SUMMARY

The present chapter offers a thorough exposition of the basic principles of latex materials, encompassing a range of topics such as latex synthesis, applications, durability, film formation, rheology, alterations, and prospective developments.

The text commences with an introductory section that expounds upon the significance and benefits of latex materials across diverse sectors. The chapter explores the various categories of latexes, encompassing both naturally occurring latexes sourced from natural origins and synthetic latexes generated via emulsion polymerization.

The text delves into the physical characteristics of latex particles, encompassing aspects such as distribution of particle sizes, shape, as well as surface chemistry, that exert an impact on their stability as well as properties.

This study delves into the formulation as well as composition of latex materials, with a particular emphasis on the polymers employed, the function of surfactants as well as emulsifiers in stabilization, as well as the integration of coalescing agents as well as film-forming additives to augment film characteristics. The chapter provides an in-depth analysis of the scientific principles that govern latex materials. These principles include colloidal stability, diffusional deterioration, radical-induced polymerization, particles nucleation mechanisms, as well as the Smith-Ewart kinetics. Additionally, the topic encompasses the various mechanisms and factors that impact the stability of latex, as well as the methods employed for assessing stability. This article elucidates the mechanisms and variables implicated in the formation of latex films, encompassing the coalescence of latex particles, the impact of temperature as well as drying parameters, and the ensuing structure and characteristics of the film.

This discourse delves into the rheology of latex systems, with emphasis on the viscoelastic properties of latexes and the impact of particle concentration as well as size. This text outlines a range of rheological measurement techniques that can be utilized to effectively characterize the behavior of latex. The subsequent chapter delves into the subject matter of latex modification and functionalization.

This includes an examination of crosslinking techniques for latex films, surface modification methods, and the process of blending as well as copolymerization of latexes in order to attain specific properties and functionalities. This paper investigates the latest advancements in latex materials, particularly in the areas of production and processing methods, the incorporation of nanotechnology in latex uses, as well as the creation of bio-based and sustainable latex materials, in order to identify emerging technologies as well as future trends.

The final section of the chapter pertains to the subject of latex recycling and its associated environmental implications. The text delves into the obstacles and approaches pertaining to latex recycling, the ecological ramifications of latex disposal, as well as the significance of adopting sustainable methodologies in latex manufacturing and utilization. The discourse briefly touches upon the environmental legislation that pertains to latex materials. In general, this chapter functions as a thorough initiation to the basics of latex materials, establishing a sturdy basis for comprehending the amalgamation, characteristics, uses, and ecological facets of latex materials.

REVIEW QUESTIONS

1. What are the different classes of latexes?
2. Explain the importance and advantages of latex materials.
3. What factors affect the stability of latex?
4. Describe the role of surfactants and emulsifiers in latex formulation.

5. How do colloidal stabilization and diffusional degradation impact latex properties?
6. Discuss the mechanisms of latex film formation and the factors influencing it.
7. What is the rheological behavior of latex systems and how is it measured?
8. Explain the concepts of latex modification and functionalization.
9. What are the emerging technologies and future trends in latex materials?
10. Discuss the challenges and strategies for latex recycling and the considerations associated with latex disposal.

REFERENCES

1. Agrawal, A. A., & Konno, K., (2009). Latex: A model for understanding mechanisms, ecology, and evolution of plant defense against herbivory. Annu. Rev. Ecol. Evol. Syst., 40, 311–331.
2. Arkaban, H., Barani, M., Akbarizadeh, M. R., Pal Singh, C. N., Jadoun, S., Dehghani, S. M., & Zarrintaj, P., (2022). Polyacrylic acid nanoplatforms: Antimicrobial, tissue engineering, and cancer theranostic applications. Polymers, 14(6), 1259.
3. Asadinezhad, A., Lehocký, M., Sáha, P., & Mozetič, M., (2012). Recent progress in surface modification of polyvinyl chloride. Materials, 5(12), 2937–2959.
4. Bano, D., Kumar, V., Singh, V. K., Chandra, S., Singh, D. K., Yadav, P. K., & Hasan, S. H., (2018). A facile and simple strategy for the synthesis of label free carbon quantum dots from the latex of Euphorbia milii and its peroxidase-mimic activity for the naked eye detection of glutathione in a human blood serum. ACS Sustainable Chemistry & Engineering, 7(2), 1923–1932.
5. Bar, H., Bhui, D. K., Sahoo, G. P., Sarkar, P., De, S. P., & Misra, A., (2009). Green synthesis of silver nanoparticles using latex of Jatropha curcas. Colloids and Surfaces A: Physicochemical and Engineering Aspects, 339(1–3), 134–139.
6. Barbosa, J. V., Moniz, J., Mendes, A., Magalhaes, F. D., & Bastos, M. M., (2014). Incorporation of an acrylic fatty acid derivative as comonomer for oxidative cure in acrylic latex. Journal of Coatings Technology and Research, 11, 765–773.
7. Barkade, S. S., Pinjari, D. V., Singh, A. K., Gogate, P. R., Naik, J. B., Sonawane, S. H., & Pandit, A. B., (2013). Ultrasound assisted miniemulsion polymerization for preparation of polypyrrole–zinc oxide (PPy/ZnO) functional latex for liquefied petroleum gas sensing. Industrial & Engineering Chemistry Research, 52(23), 7704–7712.
8. Butler, L. N., Fellows, C. M., & Gilbert, R. G., (2005). Effect of surfactants used for binder synthesis on the properties of latex paints. Progress in Organic Coatings, 53(2), 112–118.
9. Caballero, M. L., & Quirce, S., (2015). Identification and practical management of latex allergy in occupational settings. Expert Review of Clinical Immunology, 11(9), 977–992.
10. Charleux, B., D'Agosto, F., & Delaittre, G., (2010). Preparation of hybrid latex particles and core–shell particles through the use of controlled radical polymerization techniques in aqueous media. Hybrid Latex Particles: Preparation with (Mini) Emulsion Polymerization, 125–183.

11. Chern, C. S., Chen, T. J., & Liou, Y. C., (1998). Miniemulsion polymerization of styrene in the presence of a water-insoluble blue dye. Polymer, 39(16), 3767–3777.
12. Chevalier, Y., Pichot, C., Graillat, C., Joanicot, M., Wong, K., Maquet, J., & Cabane, B., (1992). Film formation with latex particles. Colloid and Polymer Science, 270, 806–821.
13. Cho, I., & Lee, K. W., (1985). Morphology of latex particles formed by poly (methyl methacrylate) - seeded emulsion polymerization of styrene. Journal of Applied Polymer Science, 30(5), 1903–1926.
14. Collyer, A. A., (1993). Techniques in Rheological Measurement. London: Chapman & Hall.
15. Das, R. K., Sharma, P., Nahar, P., & Bora, U., (2011). Synthesis of gold nanoparticles using aqueous extract of Calotropis procera latex. Materials Letters, 65(4), 610–613.
16. Datta, R. N., & BV, T. T., (2004). Environmental issues and challenges in the European latex industry. Rubber and Gum, Plastics, 57(6), 310–312.
17. Davalos-Pantoja, L., Ortega-Vinuesa, J. L., Bastos-González, D., & Hidalgo-Alvarez, R., (2001). Colloidal stability of IgG-and IgY-coated latex microspheres. Colloids and Surfaces B: Biointerfaces, 20(2), 165–175.
18. Dillon, R. E., Matheson, L. A., & Bradford, E. B., (1951). Sintering of synthetic latex particles. Journal of Colloid Science, 6(2), 108–117.
19. Dobler, F., & Holl, Y., (1996). Mechanisms of latex film formation. Trends in Polymer Science, 5(4), 145–151.
20. Einarson, M. B., & Berg, J. C., (1993). Electrosteric stabilization of colloidal latex dispersions. Journal of Colloid and Interface Science, 155(1), 165–172.
21. Eliçabe, G. E., & García-Rubio, L. H., (1989). Latex particle size distribution from turbidimetry using inversion techniques. Journal of Colloid and Interface Science, 129(1), 192–200.
22. Fernando, R. H., & Sung, L. P., (2009). Nanotechnology Applications in Coatings (Vol. 1008, pp. 2–21). Washington, DC: American Chemical Society.
23. Fujii, S., Kakigi, Y., Suzaki, M., Yusa, S. I., Muraoka, M., & Nakamura, Y., (2009). Synthesis of stimuli - responsive macroazoinitiators and their use as an inistab toward hairy polymer latex particles. Journal of Polymer Science Part A: Polymer Chemistry, 47(13), 3431–3443.

24. Galembeck, F., Braga, M., & Cardoso, A. H., (2000). Latex coagulation induced by alcohol vapors: An unusual factor for colloidal instability. Colloids and Surfaces A: Physicochemical and Engineering Aspects, 164(2, 3), 217–225.

25. Gawchik, S. M., (2011). Latex allergy. Mount Sinai Journal of Medicine: A Journal of Translational and Personalized Medicine, 78(5), 759–772.

26. Grunlan, J. C., Ma, Y., Grunlan, M. A., Gerberich, W. W., & Francis, L. F., (2001). Monodisperse latex with variable glass transition temperature and particle size for use as matrix starting material for conductive polymer composites. Polymer, 42(16), 6913–6921.

27. Guidelli, E. J., Ramos, A. P., Zaniquelli, M. E. D., & Baffa, O., (2011). Green synthesis of colloidal silver nanoparticles using natural rubber latex extracted from Hevea brasiliensis. Spectrochimica Acta Part A: Molecular and Biomolecular Spectroscopy, 82(1), 140–145.

28. Hawkett, B. S., Napper, D. H., & Gilbert, R. G., (1977). General solution to the Smith–Ewart equation for emulsion polymerization kinetics. Journal of the Chemical Society, Faraday Transactions 1: Physical Chemistry in Condensed Phases, 73, 690–698.

29. Hellgren, A. C., Weissenborn, P., & Holmberg, K., (1999). Surfactants in water-borne paints. Progress in Organic Coatings, 35(1–4), 79–87.

30. Hester, R. D., & Squire, D. R., (1997). Rheology of waterborne coatings. Journal of Coatings Technology, 68, 109–114.

31. Ho, C. C., (2021). Fundamentals and recent applications of natural rubber latex in dipping technology. Chemistry, Manufacture, and Application of Natural Rubber, 317–361.

32. Homma, A. K. O., (1992). The dynamics of extraction in Amazonia: A historical perspective. Advances in Economic Botany, 23–31.

33. Hudlikar, M., Joglekar, S., Dhaygude, M., & Kodam, K., (2012). Latex-mediated synthesis of ZnS nanoparticles: Green synthesis approach. Journal of Nanoparticle Research, 14, 1–6.

34. Jacob, J. L., D'Auzac, J., & Prevot, J. C., (1993). The composition of natural latex from Hevea Brasiliensis. Clinical Reviews in Allergy, 11, 325–337.

35. Jacob, J. L., Prévôt, J. C., Roussel, D., Lacrotte, R., Serres, E., D'Auzac, J., & Omont, H., (2018). Yield-limiting factors, latex physiological parameters, latex diagnosis, and clonal topology. In: Physiology of Rubber Tree Latex (pp. 345–382). CRC Press.

36. Jasinski, F., Lobry, E., Chemtob, A., Croutxe-Barghorn, C., Vidal, L., Josien, L., & Criqui, A., (2014). Stripping the latex: The challenge of miniemulsion polymerization without initiator, costabilizer and surfactant. Colloid and Polymer Science, 292, 3095–3102.

37. Juhue, D., & Lang, J., (1995). Film formation from dispersion of core-shell latex particles. Macromolecules, 28(4), 1306–1308.

38. Kan, C. S., & Blackson, J. H., (1996). Effect of ionomeric behavior on the viscoelastic properties and morphology of carboxylated latex films. Macromolecules, 29(21), 6853–6864.

39. Kaneko, M., Abe, T., Sekiguchi, M., Tadokoro, Y., Fukazawa, K., Yamashita, S., & Takato, S., (2010). CAS-aided visualization in LATEX documents for mathematical education. Teaching Mathematics and Computer Science, 8(1), 1–18.

40. Karaman, M. E., Meagher, L., & Pashley, R. M., (1993). Surface chemistry of emulsion polymerization. Langmuir, 9(5), 1220–1227.

41. Karp, E. M., Eaton, T. R., Sànchez, I., Nogué, V., Vorotnikov, V., Biddy, M. J., Tan, E. C., & Beckham, G. T., (2017). Renewable acrylonitrile production. Science, 358(6368), 1307–1310.

42. Keddie, J. L., Routh, A. F., Keddie, J. L., & Routh, A. F., (2010). An introduction to latex and the principles of colloidal stability. Fundamentals of Latex Film Formation: Processes and Properties, 1–26.

43. Keddie, J., & Routh, A. F., (2010). Fundamentals of Latex Film Formation: Processes and Properties. Springer Science & Business Media.

44. Keddie, J., & Routh, A. F., (2010). Fundamentals of Latex Film Formation: Processes and Properties. Springer Science & Business Media.

45. Khan, M. A., & Armes, S. P., (2000). Conducting polymer - coated latex particles. Advanced Materials, 12(9), 671–674.

46. Kheirabadi, S., & Sheikhi, A., (2022). Recent advances and challenges in recycling and reusing biomedical materials. Current Opinion in Green and Sustainable Chemistry, 100695.

47. Klass, C. P., (2007). New Nanoparticle Latex Offers Natural Advantage (Vol. 2, No. 1, pp. 30, 31). Paper360 Magazine.

48. Knauss, W. G., & Kenner, V. H., (1980). On the hygrothermomechanical characterization of polyvinyl acetate. Journal of Applied Physics, 51(10), 5131–5136.

49. Kolter, K., Dashevsky, A., Irfan, M., & Bodmeier, R., (2013). Polyvinyl acetate-based film coatings. International Journal of Pharmaceutics, 457(2), 470–479.

50. Koral, J., Ullman, R., & Eirich, F. R., (1958). The adsorption of polyvinyl acetate. The Journal of Physical Chemistry, 62(5), 541–550.
51. Kreuter, J., (1983). Physicochemical characterization of polyacrylic nanoparticles. International Journal of Pharmaceutics, 14(1), 43–58.
52. Macchini, R., Bradley, M. S. A., & Deng, T., (2013). Influence of particle size, density, particle concentration on bend erosive wear in pneumatic conveyors. Wear, 303(1, 2), 21–29.
53. Madge, E. W., Collier, H. M., & Duckworth, I. H., (1952). Mechanical stability testing of Hevea latex. Rubber Chemistry and Technology, 25(3), 630–640.
54. McDonald, C. J., & Devon, M. J., (2002). Hollow latex particles: Synthesis and applications. Advances in Colloid and Interface Science, 99(3), 181–213.
55. Min, K. E., Hu, J., Wang, C., & Elaissari, A., (2002). Surface modification of polystyrene latex particles via atom transfer radical polymerization. Journal of Polymer Science Part A: Polymer Chemistry, 40(7), 892–900.
56. Mohajerani, A., Burnett, L., Smith, J. V., Markovski, S., Rodwell, G., Rahman, M. T., & Maghool, F., (2020). Recycling waste rubber tyres in construction materials and associated environmental considerations: A review. Resources, Conservation and Recycling, 155, 104679.
57. Mondal, A. K., Mondal, S., Samanta, S., & Mallick, S., (2011). Synthesis of ecofriendly silver nanoparticle from plant latex used as an important taxonomic tool for phylogenetic interrelationship advances in bioresearch. Synthesis, 2, 31–33.
58. Mousa, A., & Karger - Kocsis, J., (2001). Rheological and thermodynamical behavior of styrene/butadiene rubber - organoclay nanocomposites. Macromolecular Materials and Engineering, 286(4), 260–266.
59. Noble, B. B., & Coote, M. L., (2013). First principles modelling of free-radical polymerization kinetics. International Reviews in Physical Chemistry, 32(3), 467–513.
60. Ortega-Vinuesa, J. L., & Bastos-González, D., (2001). A review of factors affecting the performances of latex agglutination tests. Journal of Biomaterials Science, Polymer Edition, 12(4), 379–408.
61. Paton, F. F., Peel, J. D., & Sutno, G., (1953). Factors affecting the stability of hevea latex. II. Rubber Chemistry and Technology, 26(2), 430–440.
62. Peluso, N. L., (2012). What's nature got to do with it? A situated historical perspective on socio - natural commodities. Development and Change, 43(1), 79–104.

63. Perez-Amodio, S., Holownia, P., Davey, C. L., & Price, C. P., (2001). Effects of the ionic environment, charge, and particle surface chemistry for enhancing a latex homogeneous immunoassay of C-reactive protein. Analytical Chemistry, 73(14), 3417–3425.

64. Qin, F., Li, M., Li, X. Y., & Wang, H. Q., (2015). Stability and mechanical properties of hard/soft latex blends. In: Applied Mechanics and Materials (Vol. 731, pp. 483–487). Trans Tech Publications Ltd.

65. Rajkuberan, C., Prabukumar, S., Sathishkumar, G., Wilson, A., Ravindran, K., & Sivaramakrishnan, S., (2017). Facile synthesis of silver nanoparticles using Euphorbia antiquorum L. latex extract and evaluation of their biomedical perspectives as anticancer agents. Journal of Saudi Chemical Society, 21(8), 911–919.

66. Rasutis, D., Soratana, K., McMahan, C., & Landis, A. E., (2015). A sustainability review of domestic rubber from the guayule plant. Industrial Crops and Products, 70, 383–394.

67. Ren, S., Liu, X., Fan, W., Wang, H., & Erkens, S., (2019). Rheological properties, compatibility, and storage stability of SBS latex-modified asphalt. Materials, 12(22), 3683.

68. Rogach, A., Susha, A., Caruso, F., Sukhorukov, G., Kornowski, A., Kershaw, S., & Weller, H., (2000). Nano - and microengineering: 3 - d colloidal photonic crystals prepared from sub - μm - sized polystyrene latex spheres pre - coated with luminescent polyelectrolyte/nanocrystal shells. Advanced Materials, 12(5), 333–337.

69. Routh, A. F., & Russel, W. B., (1999). A process model for latex film formation: Limiting regimes for individual driving forces. Langmuir, 15(22), 7762–7773.

70. Sadhu, S., & Bhowmick, A. K., (2004). Preparation and properties of styrene–butadiene rubber-based nanocomposites: The influence of the structural and processing parameters. Journal of Applied Polymer Science, 92(2), 698–709.

71. Samantarai, S., Nag, A., Singh, N., Dash, D., Basak, A., Nando, G. B., & Das, N. C., (2019). Chemical modification of nitrile rubber in the latex stage by functionalizing phosphorylated cardanol prepolymer: A bio-based plasticizer and a renewable resource. Journal of Elastomers & Plastics, 51(2), 99–129.

72. Scott, P. J., Kasprzak, C. R., Feller, K. D., Meenakshisundaram, V., Williams, C. B., & Long, T. E., (2020). Light and latex: Advances in the photochemistry of polymer colloids. Polymer Chemistry, 11(21), 3498–3524.

73. Serra, J., Puig, J., Martin, A., Galisteo, F., Gálvez, M., & Hidalgo-Alvarez, R., (1992). On the adsorption of IgG onto polystyrene particles: Electrophoretic mobility and critical coagulation concentration. Colloid and Polymer Science, 270, 574–583.
74. Slotved, H. C., Kaltoft, M., Skovsted, I. C., Kerrn, M. B., & Espersen, F., (2004). Simple, rapid latex agglutination test for serotyping of pneumococci (pneumotest-latex). Journal of Clinical Microbiology, 42(6), 2518–2522.
75. Smeets, N. M., Imbrogno, S., & Bloembergen, S., (2017). Carbohydrate functionalized hybrid latex particles. Carbohydrate Polymers, 173, 233–252.
76. Sperry, P. R., Snyder, B. S., O'Dowd, M., & Lesko, P. M., (1994). Role of water in particle deformation and compaction in latex film formation. Langmuir, 10(8), 2619–2628.
77. Steward, P. A., Hearn, J., & Wilkinson, M. C., (2000). An overview of polymer latex film formation and properties. Advances in Colloid and Interface Science, 86(3), 195–267.
78. Sundberg, D. C., & Durant, Y. G., (2003). Latex particle morphology, fundamental aspects: A review. Polymer Reaction Engineering, 11(3), 379–432.
79. Teo, B. M., Chen, F., Hatton, T. A., Grieser, F., & Ashokkumar, M., (2009). Novel one-pot synthesis of magnetite latex nanoparticles by ultrasound irradiation. Langmuir, 25(5), 2593–2595.
80. Thakore, S. I., Nagar, P. S., Jadeja, R. N., Thounaojam, M., Devkar, R. V., & Rathore, P. S., (2019). Sapota fruit latex mediated synthesis of Ag, Cu mono and bimetallic nanoparticles and their in vitro toxicity studies. Arabian Journal of Chemistry, 12(5), 694–700.
81. Turjanmaa, K., Kanto, M., Kautiainen, H., Reunala, T., & Palosuo, T., (2002). Long-term outcome of 160 adult patients with natural rubber latex allergy. Journal of Allergy and Clinical Immunology, 110(2), S70–S74.
82. Ugelstad, J., & Hansen, F. K., (1976). Kinetics and mechanism of emulsion polymerization. Rubber Chemistry and Technology, 49(3), 536–609.
83. Verhaar, G., (1959). Natural latex as a colloidal system. Rubber Chemistry and Technology, 32(5), 1627–1659.
84. Wall, F. T., & Gill, S. J., (1954). Interaction of cupric ions with polyacrylic acid. The Journal of Physical Chemistry, 58(12), 1128–1130.
85. Wang, Z., An, C., Lee, K., Chen, X., Zhang, B., Yin, J., & Feng, Q., (2022). Physicochemical change and microparticle release from disposable gloves in the aqueous environment impacted by accelerated weathering. Science of The Total Environment, 832, 154986.

86. Weber, R. J., Marti, J. J., McMurry, P. H., Eisele, F. L., Tanner, D. J., & Jefferson, A., (1996). Measured atmospheric new particle formation rates: Implications for nucleation mechanisms. Chemical Engineering Communications, 151(1), 53–64.
87. Winkler, D. E., (1959). Mechanism of polyvinyl chloride degradation and stabilization. Journal of Polymer Science, 35(128), 3–16.
88. Winnik, M. A., (1997). Latex film formation. Current Opinion in Colloid & Interface Science, 2(2), 192–199.
89. Winnik, M. A., (2002). Interdiffusion and crosslinking in thermoset latex films. Journal of Coatings Technology, 74(925), 49–63.
90. Wongkhat, J., & Pattamaprom, C., (2022). Enzymatic approach in accelerating colloidal stability of concentrated natural rubber latex and its effect on rubber products. Industrial Crops and Products, 182, 114892.
91. Yang, S., Castilleja, J. R., Barrera, E. V., & Lozano, K., (2004). Thermal analysis of an acrylonitrile–butadiene–styrene/SWNT composite. Polymer Degradation and Stability, 83(3), 383–388.
92. Yokoyama, A., Srinivasan, K. R., & Fogler, H. S., (1988). Stabilization mechanism of colloidal suspensions by gum tragacanth: The influence of pH on stability. Journal of Colloid and Interface Science, 126(1), 141–149.
93. Zhang, L., Wang, Y., Wang, Y., Sui, Y., & Yu, D., (2000). Morphology and mechanical properties of clay/styrene - butadiene rubber nanocomposites. Journal of Applied Polymer Science, 78(11), 1873–1878.
94. Zhao, C. L., Porzio, S., Smith, A., Ge, H., Davis, H. T., & Scriven, L. E., (2006). Direct observation of freeze-thaw instability of latex coatings via high pressure freezing and cryogenic SEM. JCT Research, 3, 109–115.
95. Zhao, Y., & Urban, M. W., (2000). Phase separation and surfactant stratification in styrene/n-butyl acrylate copolymer and latex blend films. 17. A spectroscopic study. Macromolecules, 33(6), 2184–2191.
96. Zin, W. M. B. W., Mohid, N., & Razali, M. Y., (1995). RVNRL a potential material in latex dipped products manufacturing. Radiation Physics and Chemistry, 46(4–6), 1019–1023.
97. Zosel, A., & Ley, G., (1993). Influence of crosslinking on structure, mechanical properties, and strength of latex films. Macromolecules, 26(9), 2222–2227.

CHAPTER

2 Fundamentals of Latex Chemistry

Contents

2.1. INTRODUCTION

Latex chemistry is paramount in diverse industrial sectors such as fabrics, coverings, adhesives, glues, and healthcare. Comprehending latex chemistry is imperative in developing superior products possessing desirable characteristics. The final properties of latex particles are significantly impacted by the complex interplay among initiators, monomers, surfactants, and catalysts during the process of latex polymerization. In addition, the capacity to regulate the stability of latex, the morphology of particles, and mechanisms of polymerization enable customized utilization in diverse domains. Hence, obtaining a rudimentary comprehension of latex chemistry is crucial to enhance formulation procedures and attaining the intended performance characteristics (Bilgin et al., 2016).

This chapter's objective is to comprehensively survey the key concepts that underlie the chemistry of latex. The text investigates the chemical makeup of latex particles, examining the influence of monomers and agents for crosslinking on their characteristics. The present chapter provides an in-depth analysis of diverse polymerization mechanisms, including radical, cationic, and anionic polymerization, and highlights the importance of copolymerization in latex systems. Furthermore, the chapter underscores the significance of surfactants and stabilizing agents in the stabilization of latex, specifically emphasizing their functions and prevalent categories utilized in latex chemistry. Additionally, the article examines the utilization of catalysts and initiators to regulate latex polymerization procedures. The chapter ultimately offers valuable insights into the methodologies employed for analyzing particle size and molecular weight in latex polymers, thereby facilitating the characterization and enhancement of latex properties by researchers and practitioners (Keddie et al., 2010).By acquiring a comprehensive understanding of the fundamental principles of latex chemistry, individuals can acquire significant insights to effectively navigate the intricate aspects of latex composition, processing, and utilization. This chapter aims to establish a robust groundwork for the investigation and comprehension of sophisticated latex chemistry subjects. This will enable scientists, researchers, and industry experts to create novel latex-derived commodities with improved performance and functionality (Ho, 2021).

2.2. LATEX PARTICLE COMPOSITION

Latex particles are intricate entities with diverse chemical constituents, collectively giving rise to their distinct characteristics and operational attributes. The primary constituents of latex particles are monomers, which serve as the fundamental units for the polymer chains. The monomers above engage in polymerization reactions, forming the polymer matrix that constitutes the latex particles. The selection of monomers significantly impacts the physical and chemical characteristics of latex, including elasticity, adhesion, and durability (Jacob et al., 1993). Furthermore, crosslinking agents have a pivotal function in altering the structure and characteristics of latex particles.

Incorporating these agents facilitates the formation of intermolecular bridges among the polymer chains, resulting in the development of a robust network that augments the mechanical potency and steadfastness of the latex. Including functional groups, including hydroxyl, carboxyl, or amino groups, in the composition of latex particles facilitates additional modification and functionalization. Functional groups facilitate the integration of additives, colorants, or surface modifications, thereby broadening the scope of possible applications for materials based on latex.

Comprehending the complex structure of latex fragments is imperative in customizing their attributes for particular uses, enhancing their efficiency, and attaining preferred features such as flexibility, hydrophobicity, or biocompatibility. By understanding the influence of crosslinking agents and monomers on the composition of latex particles, scholars and practitioners can effectively manage and regulate the chemical makeup of latex to satisfy the demands of various sectors, such as coverings, fabrics, medical devices, and others (Nallamilli et al., 2015).

2.2.1. Chemical Composition of Latex Particles

The chemical composition of latex particles is intricate, comprising diverse constituents which add to their distinct characteristics and conduct. Comprehending the chemical constitution of latex particles is of utmost importance in customizing their attributes and enhancing their efficacy in diverse applications.

The fundamental constituent of the latex particles is the polymer matrix, which is generated via the polymerization of monomers. Monomers are diminutive molecules that possess responsive functional groups that are proficient in undergoing polymerization reactions. Styrene, acrylates, butadiene, and vinyl acetate are frequently employed monomers in latex chemistry. The selection of monomers considerably impacts the chemical and physical properties of the resultant latex. The addition of hydrophobic monomers has been observed to augment the water contention of latex films. Furthermore, the integration of functional monomers possessing specific properties has been found to confer adhesion or integration with other materials (Zhang et al., 2011).

Tip

Remember to consider the chemical composition of latex particles when studying their properties and behavior.

Apart from the polymer matrix, latex particles can comprise crosslinking agents. Incorporating crosslinking agents is paramount in altering the architecture and characteristics of latex particles, as they facilitate the establishment of covalent linkages among polymer chains. The intermolecular connections established by crosslinks generate an intricate meshwork within the latex, augmenting its mechanical robustness, steadfastness, and longevity. Frequently employed crosslinking agents comprise divinylbenzene, melamine formaldehyde, and ethylene glycol dimethacrylate. Using crosslinking agents facilitates the generation of latex-based substances exhibiting enhanced resilience against mechanical strain, elevated temperatures, and solvent exposure (Huang et al., 2011).

Surfactants constitute a crucial constituent of the composition of latex particles. Surfactants are crucial in stabilizing latex dispersions by reducing the surface tension between the polymer particles and the surrounding medium. Lipid molecules have a hydrophilic head and a hydrophobic tail, enabling them to adsorb at the boundary between the polymer particles and the water-based medium. The adsorption process hinders particles' aggregation and merging, thereby upholding the stability of the latex dispersion. The categorization of surfactants utilized in latex chemistry is based on their charge and functional groups and includes cationic, anionic, amphoteric, and nonionic surfactants (Fujii et al., 2011).

Additional trace constituents may be found in latex particles, contingent upon the particular composition and intended utility. Latex materials can improve their performance and durability by incorporating certain additives, including UV stabilizers, antioxidants, or biocides. Furthermore, the incorporation of dyes or pigments can be utilized to provide coloration to the latex particles. The latex surface may contain functional groups, including hydroxyl, carboxyl, or amino groups, which facilitate additional modification or functioning of the particles (Papynov et al., 2013). Comprehending the chemical constitution of latex particles enables scientists and formulators to customize the characteristics of latex materials to fulfill particular application prerequisites. The manipulation of latex particle properties, including but not limited to adhesion, flexibility, water resistance, and chemical resistance, can be achieved through the precise selection of crosslinking agents, monomers, surfactants, and additives. In addition, the capacity to alter and operate the exterior of latex particles presents prospects for integrating supplementary characteristics or facilitating compatibility with alternative materials (Herzog Cardoso et al., 1999).

2.2.2. Role of Monomers in Latex Polymerization

The fundamental constituents of latex polymerization are monomers, responsible for forming the polymer matrix of latex particles by undergoing polymerization reactions. The selection of monomers is crucial in determining the chemical, physical, and mechanical features of the resultant latex product. Comprehending the function of monomers in latex polymerization is imperative to customize the attributes of latex particles and enhance their efficacy in diverse applications. The chemical structure of monomers plays a crucial role in latex polymerization. Monomers generally comprise reactive functional groups, such as acrylate, vinyl, or methacrylate, which facilitate polymerization reactions. The polymerization process is initiated by functional groups which promote the development of covalent bonds between monomers, ultimately resulting in the growth of polymer chains.

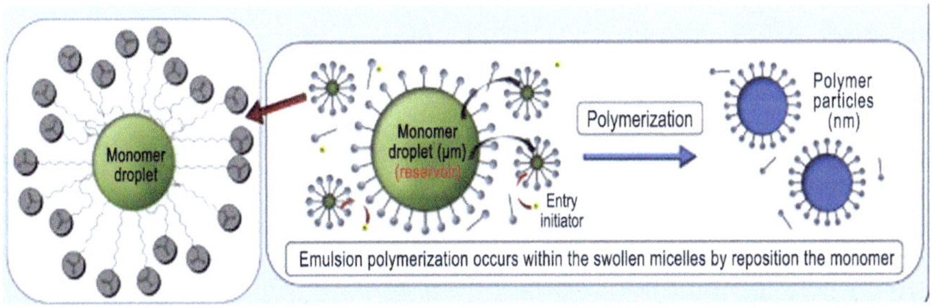

Figure 2.1. *Role of monomers in latex polymerization. The diffusion of monomers into the micelles and the micellar structure are shown as a process formed via self-organization, containing monomer droplets (head as the hydrophilic part, and the tail as the hydrophobic part) with a surfactant concentration above the critical micelle concentration are also illustrated.*

Source: A.T. Jensen, Creative Commons License.

The type and configuration of functional groups present in the constituent monomers determine the chemical properties of a polymer. These properties can include adhesion, flexibility, and chemical reactivity (Kohut et al., 2018) (Figure 2.1).

The monomers determine the physical characteristics of latex particles. The hydrophobic or hydrophilic nature of monomers impacts the latex material's surface tension and water affinity. Incorporating hydrophobic monomers, such as butadiene or styrene, into latex films augments their water resistance properties.

Conversely, including hydrophilic monomers, such as acrylic or methacrylic acid, can improve the wettability and adhesion of latex coatings. Customizing surface properties and behavior of latex particles to meet specific user requirements can be achieved by meticulously selecting and integrating particular monomers into the latex formulation (Chai et al., 2004).

The impact of monomers on the molecule weight of the polymer chains present in the latex particles is a crucial parameter to consider. The reactivity ratios of distinct monomers dictate how they assimilate into the polymer chain during the polymerization process. Copolymerization is facilitated when monomers exhibit comparable reactivity ratios, forming either block or random copolymers.

On the other hand, the utilization of monomers possessing notably distinct reactivity ratios can result in the production of copolymers that are either grafted or segmented (Gonzalez et al., 2008). Modifying monomer ratios and polymerization conditions can regulate the molecular weight distribution of the resultant latex material. In addition, the chemical properties of monomers play a significant role in the capacity for subsequent modification of latex particles.

Monomers can be deliberately chosen to incorporate distinct functional groups, such as hydroxyl, carboxyl, or amino groups, which facilitate subsequent modifications after polymerization. Incorporating functional groups in latex materials offers opportunities for crosslinking, surface modification, and chemical reactions with other materials, thereby broadening the scope of their potential applications. Furthermore, the selection of monomers can incorporate distinct characteristics, such as UV absorption, flame resistance, or conductive functionalities, within the latex polymer, thereby facilitating improved functionality or targeted uses (Khorassani et al., 2009).

2.2.3. Crosslinking Agents and Impact on Latex Properties

Introducing covalent bonds between polymer chains by crosslinking agents is crucial to latex chemistry. This process leads to the creation a 3-D network within the latex particles. Crosslinks in latex materials exert a noteworthy influence on their properties and performance. Comprehending the function of crosslinking substances and their impact on the properties of latex is imperative in customizing latex compositions to fulfill particular application prerequisites. The utilization of crosslinking agents is known to significantly improve the mechanical robustness and stability of the latex material. The formation of crosslinks generates a network architecture that confers supplementary structural robustness and resilience against deformation. As a result of this phenomenon, the elasticity, tensile strength, and durability of the latex particles are enhanced. Materials made of Latex with higher crosslink densities demonstrate enhanced mechanical strength against stressors such as stretching, abrasion, or compression, rendering them appropriate for use in applications that require durability and long-term performance (Lee, 2005) (Figure 2.2).

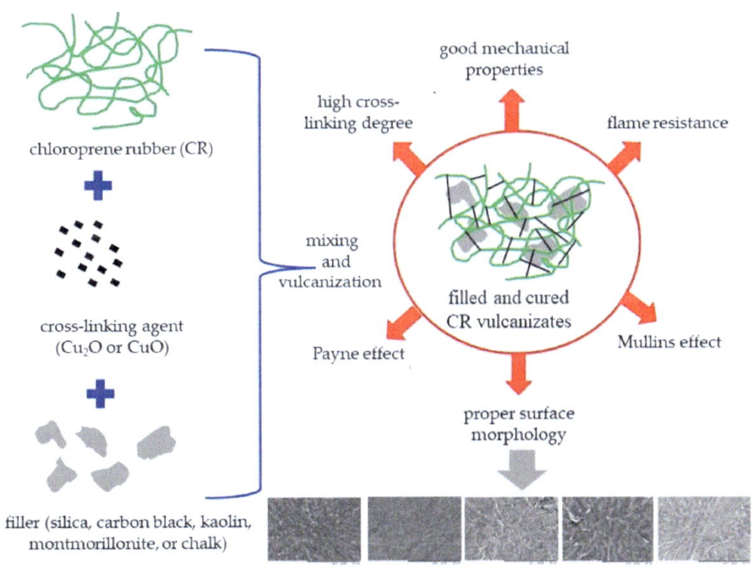

Figure 2.2. *An example of the use of copper oxides as cross-linking substances for chloroprene rubber.*

Source: Aleksandra Smejda-Krzewicka, Creative Commons License.

The existence of crosslinks also affects the thermal characteristics of latex materials. Crosslinked latex films typically demonstrate superior thermal stability and elevated temperature resistance compared to non-crosslinked latex materials. The formation of a three-dimensional network through crosslinking processes impedes the movement and diffusion of polymer chains, resulting in enhanced thermal resistance and dimension stability. Crosslinked latex characteristics render it appropriate for utilization in scenarios that necessitate high-temperature endurance, such as developing coatings for electronic devices or automotive components. The water-resistance and solvent-resistant properties of latex materials are notably affected by crosslinking agents. The incorporation of crosslinks within a polymer matrix results in a decrease in chain mobility and subsequently restricts the ingress of solvents or water molecules into the latex particles (Yan et al., 2013).

As a consequence, there is an enhancement in the capacity to resist water absorption, dissolution, and swelling. Crosslinked latex materials are frequently employed when there is a need to avoid exposure to moisture or chemicals. These applications may include coatings for outdoor constructions, glues in moist conditions, or protective films for packaging materials (Imam et al., 1999).

Moreover, the selection of crosslinking agents may impact the flexibility or rigidity of the latex substance. Crosslinking agents possessing shorter or more inflexible structures tend to induce a greater extent of crosslinking, thereby forming a more rigid latex material. On the other hand, using crosslinking agents possessing elongated or pliant structures can impart greater flexibility and suppleness to the latex substrate. The attribute of flexibility can be beneficial in scenarios requiring elasticity, traceability, or conformability.

This is particularly relevant in applications such as textile coatings or medical devices. The concentration of crosslinking agents can influence the properties of latex materials.

Elevated levels of crosslinking agents typically result in more compact crosslinking networks, which yield materials exhibiting heightened rigidity and diminished elongation at fracture. Nevertheless, an overabundance of crosslinking may result in fragility or decrease the capacity to form a film. Hence, meticulous optimization of the concentration of crosslinking agent is imperative to attain the intended equilibrium of characteristics in the latex material (Fang & Lin, 2015).

2.3. POLYMERIZATION MECHANISMS IN LATEX

The process of latex polymerization encompasses diverse mechanisms that govern the development and configuration of polymer chains inside the latex particles. The field of latex chemistry utilizes three primary polymerization mechanisms: anionic polymerization, radical polymerization, and cationic polymerization. The process of radical polymerization involves a chemical reaction by generating free radicals by a radical initiator. These free radicals then proceed to react with monomers, resulting in the formation of polymer chains. The range of uses and simplicity of initiating this mechanism has led to its widespread usage. The process of anionic polymerization entails the utilization of anionic initiators to produce highly reactive anion species that trigger the polymerization of monomers, leading to the expansion of polymer chains (Lichti et al., 1983) (Figure 2.3).

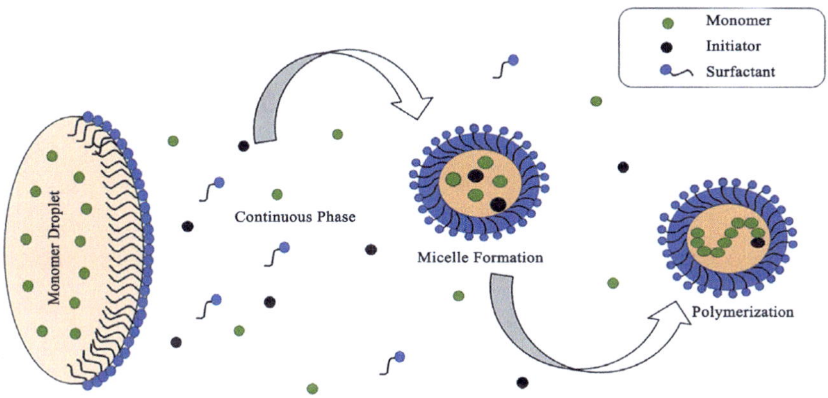

Figure 2.3. *The basic mechanisms of latex polymerization.*

Source: Sajid Iqbal, Creative Commons License.

In contrast, cationic polymerization is predicated upon the utilization of cationic initiators that produce highly reactive cationic species, which in turn initiate the polymerization of monomers. Distinct advantages and limitations are associated with each polymerization mechanism concerning reaction conditions, monomer selection, and control over polymer properties. Comprehending these mechanisms is imperative to enhance latex polymerization procedures and customize the properties of latex materials to suit particular applications (Chern, 2006).

2.3.1. Radical Polymerization

The radical polymerization mechanism is extensively employed in latex chemistry due to its adaptability, facile initiation, and capacity to assimilate diverse monomers. The polymerization process in this mechanism is triggered by the emergence of free radicals, characterized by their high reactivity and possession of unpaired electrons. The reaction is initiated by radicals that engage in a reaction with monomers, leading to the formation of polymer chains and the consequent development of the latex particles. The initiation step marks the onset of the radical polymerization process in latex. The process of initiation can be instigated either thermally or photochemically. Thermal initiation involves the decomposition of initiators, including organic peroxides or azo substances, at elevated temperatures, producing free radicals. During photochemical initiation, initiators undergo absorption of light energy, which subsequently results in the generation of free radicals. The selection of the initiator is contingent upon various factors, including the reaction temperature, the intended reaction rate, and its suitability with the monomers (Braunecker & Matyjaszewski, 2007).

Upon initiation, the process of polymerization proceeds to the propagation stage. In the propagation process, the unpaired and highly reactive molecules known as free radicals undergo a chemical reaction with monomers, leading to their assimilation into the polymer chains undergoing growth. This stage entails the iterative incorporation of monomeric units, resulting in the extension of the polymer chains. The propagation phase persists until all the monomers at disposal have been exhausted or the termination process takes place. The process of termination represents the conclusion of radical polymerization. Different mechanisms, including combination termination, can bring about termination. This involves the reaction of two polymer chains in the growth process, forming a covalent bond that ultimately halts the reaction. Disproportionation is an additional mode of termination in which two distinct free radicals engage in a chemical reaction, forming a stable molecule and the cessation of the reaction. The cessation of polymerization reactions plays a crucial role in regulating the molecular weight and the distribution of polymer chains within the latex (David et al., 2006).

The utilization of radical polymerization in latex presents various benefits. The process facilitates the integration of a broad spectrum of monomers, encompassing hydrophilic, hydrophobic, and functional monomers, thereby producing latex particles with varied characteristics. The manipulation of reaction parameters, including concentration, temperature, and reaction duration, can facilitate the regulation of the latex substance's polymerization rate and ultimate characteristics. The radical polymerization process facilitates copolymerization, allowing for the integration of diverse monomers into the polymer chains. This presents prospects for the customization of latex materials with specific properties (Goto & Fukuda, 2004).

Nevertheless, the process of radical polymerization in latex poses certain difficulties. Unintended side reactions, including chain transfer or simply termination reactions, may result in a wide range of molecular weights or the creation of low-molecular-weight entities. The existence of oxygen, impurities, or other reactive species can influence the control and stability of the radical polymerization process. Hence, meticulous screening of initiators, reaction parameters, and purification methodologies is imperative to enhance the polymerization process and attain the intended latex characteristics (Otsu, 2000).

To conclude, the mechanism of radical polymerization is an adaptable and extensively employed process in the field of latex chemistry. Incorporating a broad spectrum of monomers facilitates the production of latex particles with varied properties. The polymerization process is regulated by the beginning, transmission, and termination steps, which enable the manipulation of the molecular weight, structure, and distribution of polymer chains within the latex particles. Although there are obstacles to overcome, meticulous evaluations of initiators, conditions for reaction, and purification methods can facilitate the effective utilization of radical polymerization in the fabrication of customized latex materials for diverse applications (Wang & Matyjaszewski, 1995).

2.3.2. Anionic Polymerization

Anionic polymerization has been widely utilized in latex chemistry to produce latex particles with distinctive characteristics. The functioning of this mechanism is predicated upon the utilization of anionic initiators to generate reactive anions, which subsequently act as initiators for the polymerizing process of monomers. The process of anionic polymerization presents notable benefits, including a high degree of monomer conversion, the capacity to regulate molecular weight, and the capability to polymerize highly reactive monomers that are difficult to polymerize through alternative mechanisms. The initial stage of anionic polymerization entails the production of anionic initiators, which may comprise potent bases like alkyl metal alkoxides or alkyl lithium compounds. The initiators above facilitate the deprotonation of the monomer, thereby producing anionic entities capable of instigating the polymerization process. The selection of the initiator is contingent upon various factors, including the monomer's reactivity, the desired rate of reaction, and the stability of its initiator (Hadjichristidis et al., 2000) (Figure 2.4).

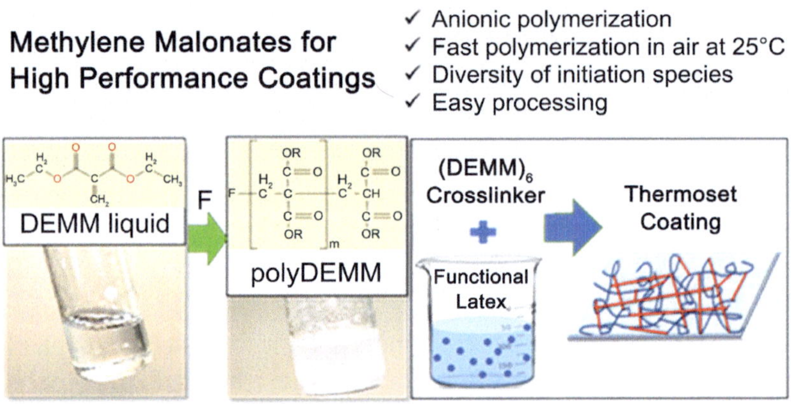

Figure 2.4. *An example of anionic polymerization of methylene malonate for high-performance coatings.*

Source: Mengfei Huang, Creative Commons License.

Upon initiation, anionic polymerization progresses via the propagation phase. In the process of propagation, anionic compounds react with monomers, leading to their assimilation into the polymer chains in the growth process. Anionic polymerization demonstrates a high degree of monomer conversion, leading to the formation of polymer structures that are precisely defined with a narrow distribution of molecular weights. Various mechanisms can lead to the termination of anionic polymerization. Coupling termination is a frequently observed mechanism of termination in polymerization, wherein two active chains undergo a chemical reaction resulting in the formation of a covalent bond, ultimately leading to the termination of the polymerization process. Additional termination modes comprise combination termination, which resembles the ones witnessed in radical polymerization. The precise regulation of termination reactions is imperative to maintain command of the molecular weight and characteristics of the latex particles (Szwarc, 2005).

The utilization of anionic polymerization presents distinct benefits. A noteworthy benefit is the capacity to polymerize monomers with high reactivity that pose difficulties for radical polymerization. Illustrative instances comprise butadiene, styrene, and isoprene. Anionic polymerization enables the production of polymers with high molecular weights and narrow molecular weight distributions, thereby ensuring their well-defined nature. Anionic polymerization is a process that offers a high degree of control of the polymer structure. This control allows for the production of block copolymers, characterized by distinct parts containing various monomers inside the polymer chain (Pantazis et al., 2003).

Nonetheless, the utilization of anionic polymerization in latex systems poses certain difficulties. The existence of contaminants, such as protic solvents, oxygen, or water, can impede the anionic polymerization. Preserving the anionic polymerization reaction's integrity mandates the implementation of specific measures, including anhydrous and oxygen-free conditions. Furthermore, it is imperative to meticulously contemplate the selection of initiators and conditions for the reaction to prevent unfavorable side reactions or early discontinuation (Morton & Fetters, 1975).

2.3.3. Cationic Polymerization

The cationic polymerization process holds considerable importance in latex chemistry as it serves as a crucial mechanism for producing latex particles that possess unique characteristics. The operational principle of this mechanism is predicated on the utilization of cationic initiators, which instigate the formation of reactive cationic species that serve as the trigger for polymerization reactions.

Cationic polymerization presents various benefits, such as the capacity to polymerize monomers with low reactivity, regulating molecular weight, and the production of polymers with distinctive structures. The cationic polymerization process commences with the production of cationic initiators during the initiation stage. Commonly employed cationic initiators are Lewis or Brønsted acids, including onium salts, Lewis acid complexes, and protonic acids.

The initiators exhibit the ability to donate protons or collaborate with the monomer, thereby engendering the formation of reactive cationic species. The selection of the initiator is contingent upon various factors, including but not limited to the monomer's reactivity, the intended reaction rate, and the stability of the initiator (Michaudel et al., 2017).

Upon initiation, the cationic polymerization process proceeds via the propagation phase. In the process of propagation, the positively charged species undergo a reaction with monomers, leading to their integration into the polymer chains in the growth process. The cationic polymerization process demonstrates a notable level of regulation over the molecular weight of polymers, thereby enabling the production of polymers with minimal variation in size distribution. Various mechanisms exist for the termination of cationic polymerization. Proton transfer termination is a frequently observed mechanism for polymer termination, whereby the polymer chain donates a proton to a monomer or another species, leading to the cessation of polymerization. An additional mode of termination entails the amalgamation of two expanding chains to generate a covalent linkage, akin to the process of radical polymerization. Precise management of cessation reactions is imperative in attaining the latex particles' intended molecular weight and characteristics (Aoshima & Kanaoka, 2009) (Figure 2.5).

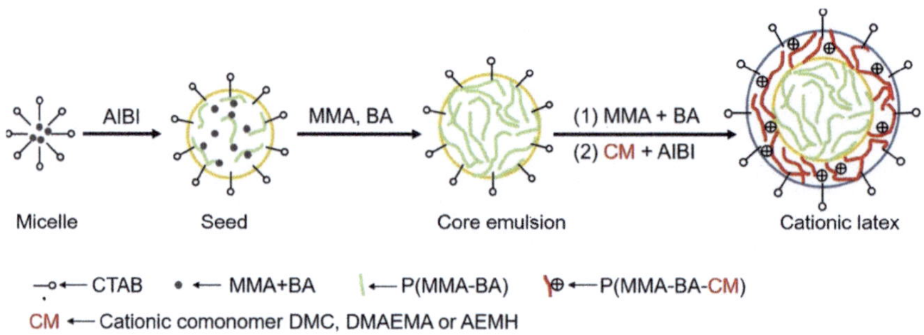

Figure 2.5. *An example of preparation, stability, and film properties of cationic polyacrylate latex particles with various substituents on the nitrogen atom.*

Source: Chen Hua, Creative Commons License.

The utilization of cationic polymerization in latex chemistry presents distinct benefits. A notable benefit is the capability to initiate the polymerization of monomers possessing low reactivity, including but not limited to cyclic ethers, vinyl ethers, and isobutylene. The restricted reactivity of these monomers in alternative polymerization mechanisms endows cationic polymerization with significant utility in their synthesis. Cationic polymerization facilitates the regulation of molecular weight and the synthesis of polymers exhibiting low polydispersity. Utilizing this control mechanism facilitates the production of polymers with accurately determined chain lengths and customized characteristics (Yağci & Reetz,

1998). Nonetheless, the field of latex chemistry encounters difficulties with cationic polymerization. The interference of impurities, such as protic solvents or moisture, can hinder the cationic polymerization process, resulting in side reactions or premature termination. To achieve successful cationic polymerization, taking special precautions, including maintaining anhydrous and inert conditions, is imperative. Furthermore, it is imperative to meticulously contemplate the selection of initiators and reaction conditions to prevent unfavorable side reactions or instability (Dworak et al., 1995).

2.3.4. Copolymerization in Latex Systems

The copolymerization process is an essential component of latex chemistry, wherein the concurrent polymerization of two or more distinct monomers occurs. The present mechanism facilitates the production of copolymers, essentially polymers that comprise two or more unique monomer units inside the polymer chain. Utilizing co-polymerization in latex systems presents numerous benefits, such as the capacity to customize properties, incorporate novel functionalities, and elevate the efficacy of the resultant latex materials. The copolymerization process commences by carefully choosing monomers that exhibit distinct chemical arrangements or properties. The selection of monomers can be based on their desired properties, including but not limited to hydrophilicity, hydrophobicity, reactivity, or functionality. The amalgamation of monomers exhibiting distinct properties enables the integration of distinctive attributes into the copolymer, presenting a diverse array of prospects for attaining particular material properties (Marcu et al., 2003).

The copolymerization process follows a sequence of steps: initiation, propagation, and termination, which is analogous to homopolymerization. In copolymerization, the reaction mixture contains both monomers and the polymer chain undergoes simultaneous incorporation of the monomers. The determination of copolymer composition and structure is significantly influenced by the reactivity ratios of the monomers, which serve as indicators of their respective reactivity towards one another. The copolymer compositions can vary depending on the reactivity ratios of the monomers and their respective combinations, leading to the formation of distinct copolymer types, including but not limited to random copolymers, block copolymers, and graft copolymers (Guillaume et al., 1988).

Random copolymers are produced due to comparable reactivity ratios of the constituent monomers, resulting in a nearly homogeneous dispersion of monomeric units throughout the polymer chain. Consequently, the copolymers exhibit a combination of characteristics from the distinct monomers. Block copolymers are produced due to the distinct reactivity ratios of the monomers, leading to the formation of discrete blocks of each monomer within the copolymer chain. This facilitates the generation of copolymers featuring distinct domains that manifest distinctive characteristics. The formation of graft copolymers entails the polymerization of one monomer as the primary backbone and the polymerization of another monomer as the secondary side chains, resulting in copolymers exhibiting distinctive architectures (Dimonie et al., 1984).

Using copolymerization to introduce diverse monomers presents many benefits in latex systems. The copolymerization process enables precise control over various properties, including but not limited to hardness, flexibility, adhesion, and compatibility with diverse materials. Through the careful selection of monomers and precise control of copolymerization parameters, it is feasible to attain latex materials that exhibit enhanced mechanical robustness, heightened thermal stability, superior chemical resistance, or targeted functionalities. The phenomenon of copolymers displaying synergistic effects is well-documented in the literature. By combining the advantageous characteristics of individual monomers, copolymers can demonstrate superior performance relative to homopolymers. In addition, the copolymerization process within latex systems presents prospects for creating materials possessing customized surface characteristics. The surface wetting of latex particles can be modulated by including either hydrophobic or hydrophilic monomers, thereby providing the ability to regulate properties such as adherence, spreading, or repulsion. This feature is especially advantageous in scenarios where surface interactions are crucial, such as coverings, adhesives, or biomedical materials (Thomas et al., 2000).

2.4. LATEX PARTICLE FORMATION AND MORPHOLOGY

The configuration and structure of latex particles are paramount in ascertaining latex substances' chemical and physical properties. Latex particle formation is achieved via emulsion polymerization, which involves the dispersion of monomers in an aqueous medium facilitated by surfactants or emulsifiers. The process of polymerization occurs within the dispersed droplets, leading to the creation of latex particles (Cho & Lee, 1985). The dimensions and morphology of the particles are contingent upon a range of factors, encompassing the concentration of the monomer, the type of emulsifier employed, the conditions of polymerization, and the degree of agitation. The morphology of latex particles exhibits a broad spectrum of shapes, ranging from spherical to irregular in shape forms, including elliptical or irregular structures. The dimensions and structure of latex particles significantly impact various characteristics, including surface area, dispersibility, film-forming ability, and mechanical robustness. As a result, these factors assume crucial significance in the formulation and processing of latex (Cal et al., 1990) (Figure 2.6).

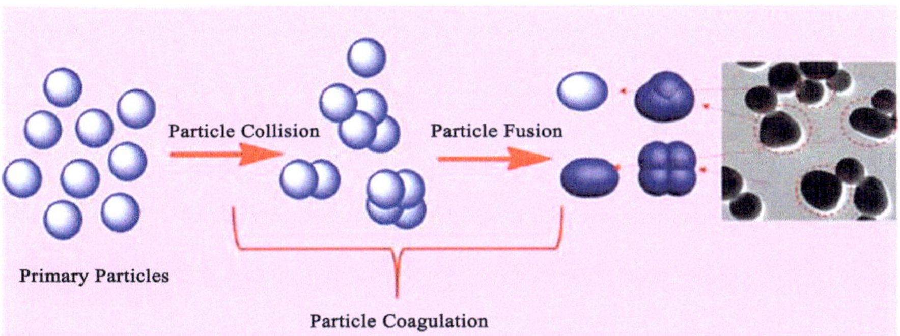

The process of cross-linking latex particle coagulation.

Figure 2.6. *The process of cross-linking latex particle coagulation.*

Source: Baijun Liu, Creative Commons License.

2.4.1. Nucleation and Coagulation Processes in Latex Formation

Emulsion polymerization entails several crucial mechanisms, such as nucleation and coagulation that contribute to developing latex particles. The processes above are paramount in ascertaining the latex particles' dimensions, dispersion, and steadfastness (Goodwin et al., 1973).

The process of latex formation commences with nucleation, which is the primary stage where minute clusters of monomers merge to generate the initial nuclci or primary particles. Two primary mechanisms for nucleation are homogeneous and heterogeneous nucleation. Homogeneous nucleation is when monomers within an emulsion polymerization system spontaneously aggregate and generate nuclei without external particles or surfaces. In contrast, heterogeneous nucleation is characterized by external particles or the surfaces, such as contaminants or latex seed particles from a prior batch, that serve as nucleation sites for accumulating monomers and forming nuclei. The nucleation process is subject to the influence of various factors, including but not limited to the concentration of monomers, temperature, type of surfactant, and emulsification conditions. The regulation of the nucleation process holds paramount importance in attaining homogenous particle size and dispersion in latex (Feeney et al., 1985) (Figure 2.7).

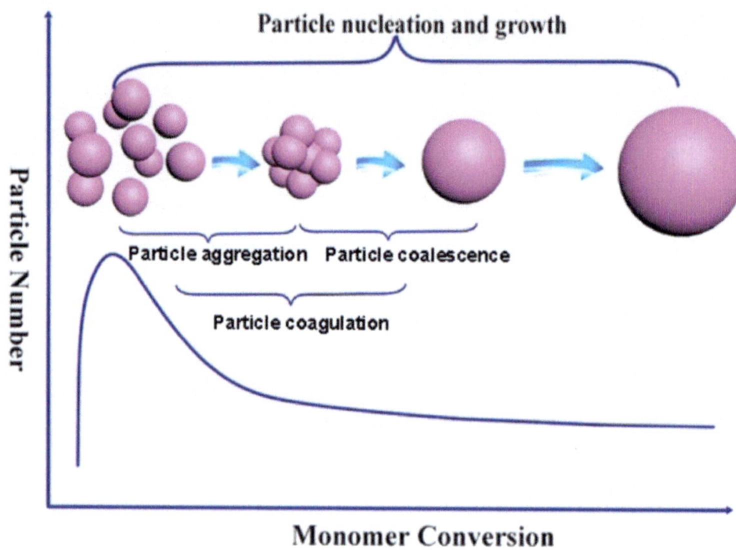

Figure 2.7. *The schematic diagram of particle coagulation in one-step emulsion polymerization.*

Source: Baijun Liu, Creative Commons License.

The phenomenon of coagulation, which is alternatively referred to as agglomeration or flocculation, pertains to the amalgamation of latex particles resulting in comparatively larger builds or clusters. The coagulation phenomenon can be attributed to many factors, such as inadequate stabilization of the latex dispersion, alterations in pH or ionic strength, or destabilizing agents. The coagulation process can be categorized into two distinct types, reversible and permanent coagulation. The process of reversible coagulation entails the creation of relaxed conglomerates that can be readily dispersed once the destabilizing factor is eliminated. Conversely, irreversible coagulation results in the creation of permanent aggregates incapable of being re-dispersed. The coagulation process can considerably impact various latex dispersion characteristics, including its stability, particle size distribution, and rheological properties (Liu et al., 2016).

Surfactants or emulsifiers are frequently employed to regulate the nucleation and coagulation mechanisms during latex formation. The latex dispersion is stabilized by utilizing surface-active agents, which function by diminishing the interfacial tension between the monomer droplets and the water-based medium. Surfactants or emulsifiers are known to adsorb at the interfaces of droplets, thereby creating a protective layer that effectively inhibits droplet coalescence and facilitates consistent nucleation and dispersion. The nucleation and coagulation procedures and the latex dispersion stability can be significantly impacted by the surfactants or emulsifiers' type, concentration, and hydrophilic-lipophilic balance (HLB) (Feeney et al., 1987).

Various factors can influence the kinetics of nucleation and coagulation processes. The nucleation process is significantly influenced by temperature, whereby higher temperatures are typically associated with elevated nucleation rates. The rate of nucleation and coagulation is influenced by the concentration of monomers, whereby an increase in monomer concentration can expedite the process of creating particles and their aggregation. The selection of a surfactant or emulsifier can considerably influence the stability and the kinetics of nucleation and coagulation. This is due to the varying affinities of different surfactants towards monomer droplets and their unique capacity to stabilize latex dispersion (Castellanos et al., 2023).

Reaching consistent particle size, narrowed size distribution, and high stability in latex formation requires proper oversight of the nucleation and coagulation processes. Comprehending the variables that impact said processes facilitates the enhancement of emulsion polymerization circumstances, surfactant choice, and formulation parameters. The precise regulation of nucleation and coagulation processes enables the production of latex materials with specific particle size, shape, and stability properties, enhancing their utility across diverse industrial sectors, including protective coatings, glues, textiles, and biomedical applications (Ukrainczyk & Rogina, 2013).

2.4.2. Emulsion Polymerization Techniques for Latex Production

The emulsion polymerization technique is an adaptable technique employed for fabricating latex materials with customized properties. Several emulsion polymerization techniques have been devised to regulate the latex's particle size, morphology, and polymer properties. Utilizing these techniques provides a degree of adaptability in creating latex materials tailored to particular applications. This discourse examines prevalent emulsion polymerization methodologies utilized in latex production (Ceska, 1974) (Figure 2.8).

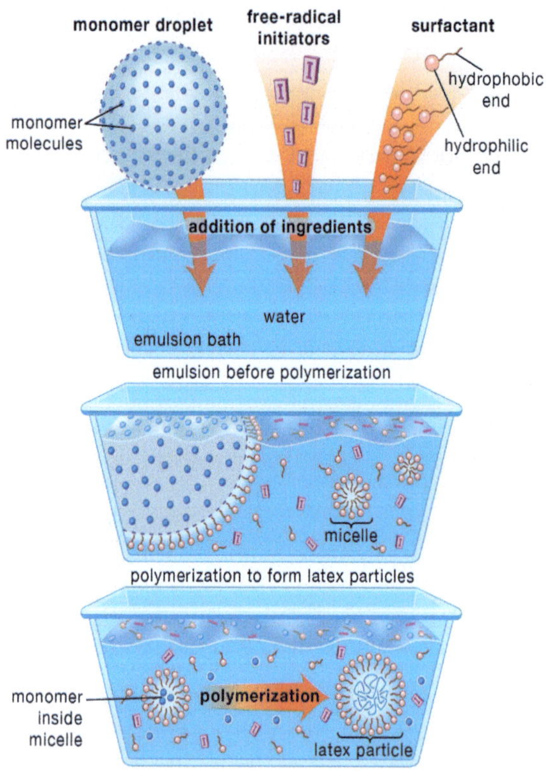

Figure 2.8. *Schematic diagram of the emulsion-polymerization method.*

Source: Britannica, Creative Commons License.

2.4.2.1. Batch Emulsion Polymerization

The most uncomplicated and direct method for polymerization through batches is emulsion polymerization. The process entails introducing initiators, monomers, and surfactants into a reactor vessel and performing the polymerization reaction while maintaining regulated conditions. The process of batch polymerization provides a means of regulating reaction variables such as monomer concentration, temperature, and reaction duration with a high degree of accuracy. The method is appropriate for generating limited amounts of latex substances that exhibit a diverse array of characteristics (Manguian et al., 2006).

2.4.2.2. Semi-batch Emulsion Polymerization

The semi-batch emulsion polymerization process is a variant of the batch methodology. The present methodology involves gradually adding monomers into the reaction vessel while ensuring a consistent reaction temperature. The implementation of a regulated monomer addition process enables enhanced management of particle dimensions, particle size variability, and polymer architecture. Semi-batch polymerization of emulsions is a common practice in producing latex materials characterized by narrowed particle size distributions and targeted molecular weight profiles (Gabriel et al., 2022).

2.4.2.3. Continuous Emulsion Polymerization

Continuous emulsion polymerization involves the continuous addition of monomers, initiators, and surfactants into a reactor, while simultaneously removing the latex product. This technique offers continuous and steady-state production of latex materials.

The methodology above provides a consistent and uninterrupted generation of latex substances. Continuous polymerization methodology enables enhanced regulation of reaction parameters, leading to superior uniformity of the resultant product and increased production rates. Large-scale industrial production is particularly advantageous (Rawlings & Ray, 1988).

2.4.2.4. Seeded Emulsion Polymerization

The process of seeded emulsion polymerization entails the introduction of pre-existing polymer particles, commonly called seeds, into the reaction medium.

The seeds above function as nucleation centers for the polymerization process, thereby aiding in creating uniform particles. The seeded emulsion polymerization technique enables enhanced regulation of particle dimensions, morphology, and polymer attributes. This technique is prevalent in the production of latex materials, enabling the attainment of narrowed distributions of particle sizes and precise particle morphologies (Mock et al., 2006).

2.4.2.5. Miniemulsion Polymerization

The miniemulsion polymerization process is a specific methodology that yields particles of minute dimensions, generally falling within the submicron scale. The process entails the dispersion of monomers within a continuous phase through high-shear mixing, ultimately forming mini-emulsion droplets. Subsequently, the commencement of polymerization takes place within the aforementioned diminutive droplets. The miniemulsion polymerization process enables the production of latex materials that exhibit remarkable stability, elevated solids content, and distinct characteristics. This material exhibits exceptional properties that make it well-suited for applications that demand outstanding durability coatings, glues, or nanocomposites (Landfester, 2009).

2.4.2.6. Surfactant-Free Emulsion Polymerization

The technique of surfactant-free emulsion polymerization involves the absence of surfactants or emulsifiers to stabilize latex dispersion. Rather than using general monomers, particular ones are chosen for their intrinsic emulsifying characteristics. The monomers undergo self-emulsification during the polymerization process, resulting in the creation of steady latex particles. Using surfactant-free emulsion polymerization presents benefits, including enhanced biocompatibility, diminished leftover surfactants, and streamlined formulation.

Each emulsion polymerization technique offers distinct advantages, making them suitable for particular applications. The selection of a particular methodology is contingent upon various aspects, including the targeted size of particles, size distribution, polymer architecture, manufacturing capacity, and ultimate application specifications. By meticulously selecting the suitable emulsion polymerization methodology, scholars and practitioners can generate latex substances with customized characteristics, facilitating their utilization across diverse industrial sectors (Hearn et al., 1985).

2.4.3. Morphology and Structure of Latex Particles

The shape and arrangement of latex particles are significant factors that influence latex materials' chemical and physical features. Comprehension and manipulation of these attributes are imperative in customizing latex compositions to fulfill particular application prerequisites. In this study, an in-depth analysis is conducted on the shape and form of latex particles (Muroi et al., 1984).

2.4.3.1. Morphology

Latex particles display a heterogeneous array of morphologies, encompassing a spectrum of shapes spanning around to non-spherical configurations. The morphology of latex particles is subject to the influence of multiple factors, such as the emulsion polymerization technique employed, the composition of the monomer, the conditions of polymerization, and the presence of any additives. The physical characteristics of particles significantly impact their properties, including but not limited to dispersibility, surface area, film formation, and mechanical durability. The prevalent morphology detected in latex systems is that of spherical particles. The particles exhibit a distinct spherical morphology and offer homogeneous coverage upon their application as coverings or films. Under certain circumstances, particles can adopt non-spherical shapes, such as ellipsoidal or erratic configurations. The particles above have the potential to display heightened flow characteristics, superior packing efficiency, or distinctive optical properties, rendering them appropriate for specific applications (Stubbs & Sundberg, 2008).

2.4.3.2. Size and Size Distribution

The dimensions and distribution of the latex particles are fundamental variables that significantly influence the efficacy and robustness of latex-based materials. Particle size is commonly assessed by determining the mean particle diameter, which can be expressed as the weight-average diameter (Dw), number-average diameter (Dn), or volume-average diameter (Dv). The size distribution pertains to the spectrum of particle sizes observable within the latex dispersion.

The precise regulation of emulsion polymerization parameters, such as emulsifier type, monomer concentration, reaction temperature, and agitation, is instrumental in controlling the particle size and size distribution. Reduced particle dimensions present benefits such as augmented interfacial area, superior coating development, and heightened obstructive characteristics. The attainment of narrow size distributions is crucial in achieving uniformity in coating thickness and minimizing the probability of particle aggregation or sedimentation (Vega et al., 2003).

2.4.3.3. Core-Shell Structure

The particles of Latex have the potential to demonstrate a core-shell configuration, wherein an outer layer encompasses a well-defined central region. The attainment of the core-shell architecture is accomplished through copolymerization or simply sequential polymerization methodologies. The central region of the structure is typically composed of a singular polymer variety, whereas the outer layer is composed of either a distinct polymer or a blend of polymers. A core-shell structure confers distinct characteristics, such as heightened mechanical robustness, regulated liberation of enclosed substances, or augmented congruity with supplementary constituents in a composition. Tailoring the core-shell structure is feasible to fulfill particular application demands. The utilization of a hydrophobic core and hydrophilic shell has been observed to enhance stability in aqueous systems. Similarly, an elastomeric shell and a thermoplastic core have been found to provide increased flexibility and impact resistance. The shell layer's thickness and composition can be manipulated to regulate the barrier characteristics, adherence, or integration of the latex particles (Dingenouts et al., 1998).

2.4.3.4. Porosity

Porosity pertains to empty spaces or cavities inside the latex particles. The formation of porous latex particles is achieved by incorporating porogens or implementing specialized polymerization methodologies. The impact of latex particle porosity on various characteristics, including but not limited to permeability, water absorption, and loading capacity for enclosed materials, has been observed.

Porous particles possess certain benefits, including but not limited to enhanced breathability, augmented area of contact for chemical reactions, and improved absorption capabilities. The manipulation of formulation or incorporation of porogenic agents throughout polymerization can regulate the porosity of latex particles. The porosity and functionality of latex particles are influenced by the dimensions, distribution, and connection of their pores. The characteristic above holds significant significance in various applications, including medication delivery systems, adsorbents, and catalyst supports (Pelley & Tufenkji, 2008).

2.4.3.5. Surface Functionality

Latex particles can be functionalized by incorporating particular monomers or post-polymerization modification methods to alter their surface properties. The surface of particles can be modified by incorporating functional groups, such as hydroxyl, carboxyl, or amino groups, which facilitate additional chemical reactions, crosslinking, or the conjugation of bioactive molecules. The surface functionality of latex particles enables their customization for various purposes, including but not limited to enhancing adhesion, modifying surface properties, or facilitating targeted delivery of pharmaceutical agents (Santos & Forcada, 1996).

2.5. SURFACTANTS AND STABILIZERS IN LATEX CHEMISTRY

The utilization of surfactants and stabilizers is crucial in latex chemistry as they aid in settling the latex dispersion, hinder coalescence, and promote emulsion polymerization. Surfactants facilitate the reduction of interfacial tension between monomer droplets and the water phase, thereby enabling the creation and stabilization of minute latex particles.

The particles adhere to the interfaces of the droplets, creating a layer of protection that impedes the merging of droplets and sustains the dispersion of particles. In contrast, stabilizers offer supplementary safeguards against flocculation and coagulation through electrostatic or steric means. These substances aid in preserving the stability and prevention of the accumulation of latex particles throughout the processes of polymerization, storage, and application. The appropriate choice of surfactants and stabilizers is paramount, as they can impact various characteristics of latex materials, including the size of particles, distribution of particle sizes, stability, rheological properties, and overall efficacy (Voorn et al., 2006) (Figure 2.9).

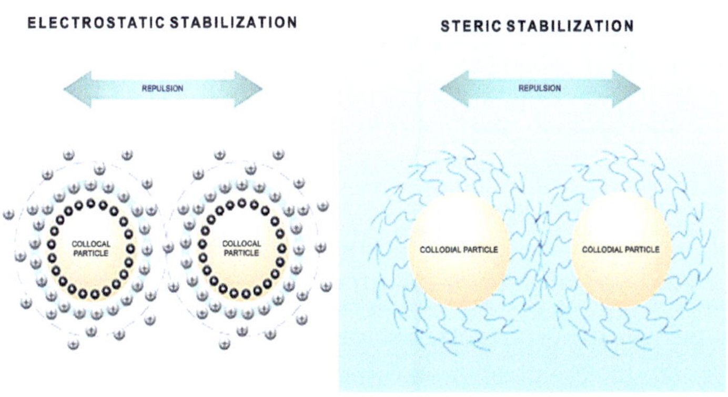

Figure 2.9. *Electrostatic and steric stabilization in latex chemistry.*

Source: MCP Polymers, Creative Commons License.

2.5.1. Role of Surfactants in Latex Stabilization

Surfactants are essential in stabilizing latex dispersions, as they facilitate creating and preserving minute, dispersed latex particles. The distinctive characteristics of these substances facilitate a reduction in interfacial tension and the prevention of coalescence, ultimately improving the stability and efficacy of latex-based materials. This section will delve into a more detailed examination of the function of surfactants in latex stabilization (Mohamed et al., 2016).

2.5.1.1. Interfacial Tension Reduction

Surfactants exhibit amphiphilic characteristics, indicating the presence of both hydrophilic (affinity towards water) and hydrophobic (repulsion towards water) moieties.

Upon introduction to a latex structure, surfactants undergo adsorption at the interface between oil and water, thereby mitigating the interfacial tension between the monomer drops and the aqueous phase. The decrease in interfacial tension facilitates the creation and maintenance of diminutive latex particles by impeding the merging of droplets (Vignati et al., 2003).

2.5.1.2. Emulsification and Dispersion

Surfactants facilitate the process of emulsification and distribution of monomers in the aqueous phase in the context of emulsion polymerization. Surfactants hinder the coalescence of monomer droplets and facilitate their homogeneous dispersion throughout the medium by creating a barrier of protection at the droplet interface. The process guarantees the production of latex particles of optimal dimensions, exhibiting a precisely regulated size and limited size variance.

2.5.1.3. Prevention of Coalescence

The principal role of surfactants in stabilizing latex is to impede coalescence, which refers to the amalgamation of neighboring latex particles. Surfactant molecules create a steric hindrance by encircling the latex particles, obstructing their direct contact.

Implementing this barrier results in a notable decrease in the likelihood of particle coalescence, thereby preserving the stability of the latex dispersion (Ata et al., 2010).

2.5.1.4. Redispersion of Agglomerates

Surfactants are also redispersing latex clusters that may arise during storing or handling. Agglomerates refer to the formation of larger components of latex particles, which can be attributed to various factors such as fluctuations in temperature, changes in pH levels, or the presence of contaminants. Using surfactants possessing suitable characteristics can facilitate the process of redispersing agglomerates, disintegrating them into individual latex particles, and reinstating the stability of the dispersion.

2.5.1.5. Electrostatic and Steric Stabilization

Surfactants offer stabilization via electrostatic and steric techniques. Electrostatic stabilization is a phenomenon that arises due to the presence of surfactant molecules that possess a net charge. This leads to the emergence of repellent forces among the charged latex particles. The phenomenon of repulsion plays a crucial role in inhibiting particle aggregation and facilitating dispersion stability. Steric stabilization is a phenomenon whereby surfactant molecules create a physical barrier surrounding the latex particles. This barrier impedes close contact and facilitates particle dispersion via steric hindrance (Fuchs & Killmann, 2001).

2.5.1.6. pH and Temperature Sensitivity

Certain surfactants demonstrate sensitivity to pH and temperature, enabling the regulation of latex stabilization under particular circumstances. The adsorption behavior of surfactants at the interfaces of latex particles can be altered by changes in their conformation or charge density, which changes in pH or temperature can induce. The responsiveness of pH or temperature can be utilized to customize the stability of latex dispersions to suit diverse processing conditions or applications. The identification of suitable surfactants is contingent upon various factors, such as the characteristics of the monomers, the targeted particle dimensions, the requisite stability parameters, and the proposed utilization of the latex substance. When selecting surfactants, it is important to consider various factors such as their compatibility with the monomer system, HLB, adsorption effectiveness, and long-term stability. Various surfactant categories, namely anionic, cationic, nonionic, and zwitterionic surfactants, present distinct benefits and are appropriate for diverse latex compositions and specifications (Deng et al., 1996).

2.5.2. Types of Surfactants Used in Latex Chemistry

A diverse range of surfactants is utilized in latex chemistry to stabilize latex dispersions and regulate their characteristics. The surfactants exhibit varying chemical compositions, charge properties, and HLB values, enabling their utilization across diverse applications and performance specifications. The subsequent passages furnish a synopsis of surfactant categories that are frequently employed in the domain of latex chemistry (Aramendia et al., 2003).

2.5.2.1. Anionic Surfactants

Anionic surfactants are a class of surfactant molecules that bear a negative charge. These surfactants are typically obtained from sulfonic, carboxylic, or phosphoric acids. Anionic surfactants, such as sodium lauryl sulfate (SLS), sodium dodecyl sulfate (SDS), and sodium alkylbenzenesulfonates, are commonly used in various applications. The efficacy of these surfactants lies in their ability to stabilize latex dispersions via electrostatic repulsion, which arises from the interaction between anionic surfactant molecules and negatively charged latex particles. Due to their favorable emulsification and dispersion characteristics, anionic surfactants are frequently employed in emulsion polymerization procedures.

2.5.2.2. Cationic Surfactants

Cationic surfactants possess a positively charged moiety and are frequently synthesized from quaternary ammonium compounds. In latex chemistry, they are employed for particular purposes that necessitate cationic stabilization or engagement with surfaces that bear negative charges. Cationic surfactants, such as cetylpyridinium chloride (CPC) and cetyltrimethylammonium bromide (CTAB), are commonly employed in latex formulations. Cationic surfactants can offer efficient stabilization through electrostatic repulsion, which arises from the interaction between the positive-charged surfactant molecules and the oppositely charged latex particles (Gan & Ng, 1986).

2.5.2.3. Nonionic Surfactants

Nonionic surfactants are distinguished by a hydrophilic headgroup and a hydrophobic tail without carrying any net charge. These surfactants exhibit versatility and are extensively employed in latex chemistry because they are compatible with diverse monomer systems and meet stability requirements.

Nonionic surfactants encompass a range of chemical compounds, such as ethoxylated fatty alcohols, ethoxylated alkylphenols, and polyethylene glycol derivatives. Nonionic surfactants exhibit remarkable emulsification characteristics and have the potential to impart steric stabilization through the creation of a physical enclosure surrounding the latex particles.

2.5.2.4. Amphiphilic Block Copolymers

Amphiphilic block copolymers are composed of multiple blocks that exhibit varying hydrophilic and hydrophobic properties. These substances possess the characteristics of both surfactants and polymers, exhibiting emulsification and stabilization functionalities. Typical instances comprise copolymers such as polyethylene oxide-block-polypropylene oxide-block-polyethylene oxide and polystyrene-block-polyethylene oxide. The block copolymers exhibit self-assembly characteristics at the interfaces of latex particles, leading to the formation of a barrier of protection that stabilizes the dispersion of latex. This stabilization is achieved through steric obstacles and interactions among the polymer sections and the latex particles (Qiao et al., 2015).

2.5.2.5. Zwitterionic Surfactants

Zwitterionic surfactants are characterized by cationic and anionic groups within a single molecule, giving rise to a charge balance and an overall neutral charge. These materials exhibit exceptional compatibility with diverse systems, favorable biocompatibility, and minimal toxicity. Betaines and sulfobetaines are among the zwitterionic surfactants commonly employed in latex chemistry. Zwitterionic surfactants can provide electrostatic and steric stability properties, making them a potentially efficacious option for stabilizing latex dispersions. The choice of surfactant category is contingent upon the intended characteristics of the latex substance, encompassing regulation of particle size, stability prerequisites, and coherence with the monomer system. Considering HLB value, surfactant concentration, and compatibility with other formulation components is imperative. A blend of surfactant varieties may be utilized in certain instances to attain the intended stability and operational attributes (Abele et al., 1999).

2.5.3. Stabilizers and Their Function in Latex Dispersions

Stabilizers are of paramount importance in the preservation of stability and the prevention of coagulation of latex dispersions. Additives, referred to as protective colloids or stabilizing agents, improve latex materials' efficacy and longevity by impeding particle aggregation and upholding a consistent dispersion. This section will delve deeper into the role of stabilizers in latex dispersions.

2.5.3.1. Electrostatic Stabilization

The fundamental role of stabilizers is to furnish electrostatic stabilization. Stabilizers possessing a suitable charge, namely cationic or anionic stabilizers, induce electrostatic repulsion among the latex particles, impeding their proximity and aggregation. The phenomenon of repulsion arises due to charged moieties on the stabilizing molecules. These charged groups engender a charged stratum encircling the latex particles, preserving their dispersion stability via electrostatic repulsion (Horák & Benedyk, 2004).

2.5.3.2. Steric Stabilization

Stabilizers can provide steric stabilization, especially nonionic or amphiphilic block copolymer stabilizers. The stabilizers exhibit adsorption onto the surfaces of latex particles, thereby creating a material barrier or "steric hindrance" that impedes the proximity of particles. The stabilizer's polymer chains function to form a protective layer surrounding the latex particles, thereby impeding particle aggregation and sustaining dispersion stability via steric repulsion.

2.5.3.3. Reducing Interparticle Forces

Stabilizing agents function by mitigating the interparticle attractive forces that might otherwise result in particles' accumulation or coalescence. The intermolecular forces between particles comprise van der Waals and fluid bridging forces. Stabilizers function by adsorbing onto the surfaces of latex particles, generating repellent forces or obstacles that combat the attractive forces. This mechanism prevents particle aggregation and preserves the stability of the dispersion (Ikeda & Nishinari, 2000).

2.5.3.4. Dispersion Rheology

Stabilizing agents have the potential to exert an impact on the rheological characteristics of latex dispersions. They play a role in determining the dispersion's viscosity, shear-thinning properties, and flow characteristics. The process of stabilizer adsorption onto the surfaces of latex particles results in the formation of a mechanical structure that impacts the viscosity of the dispersion and its behavior under the influence of shear forces. A dispersion's application and processing features are influenced by its rheological properties, including coatings, film growth, and flow behavior during mixing or pumping.

2.5.3.5. pH and Ionic Strength Stability

Some stabilizers display sensitivity to pH or ionic strength, which enables the regulation of the stability of latex fragments under particular circumstances. Variations in pH or the ions' strength can impact stabilizers' adsorption and efficacy. Conformational changes in stabilizers can modify their adsorption actions and stabilization properties. The responsiveness of pH or ionic strength can be employed to customize the stability of latex dispersions to suit particular processing conditions or applications (Garcia-Garcia et al., 2009).

2.5.3.6. Compatibility and Compatibility with Formulation Components

The compatibility of stabilizers with other formulation components, such as surfactants, monomers, crosslinking agents, and additives, is imperative. Any external factors mustn't impede the process of polymerization, the formation of latex particles, or the intended characteristics of the ultimate latex product. The optimal choice of stabilizers is contingent upon their congruity with the particular monomeric system, surfactants, and additional components of the formulation. The selection of a stabilizer is contingent upon various factors, such as the desired level of stability, the suitability of the stabilizer with the monomer system, the size of the latex particles, and the intended use of the latex material. Frequently employed stabilizers comprise polyvinyl alcohols, cellulose derivatives, hydroxyethyl cellulose, and polyvinylpyrrolidone. The effective stabilization of latex particles is contingent upon critical factors such as the concentration and composition of the stabilizer, in addition to the adsorption and interaction between the stabilizer and the particles (Kunkel & Urban, 1993).

2.6. INITIATORS AND CATALYSTS IN LATEX POLYMERIZATION

Initiators and catalysts play a crucial role in the process of latex polymerization by starting and enabling the polymerization reaction. Initiators are chemical substances that instigate chemical reactions by producing free radicals or initiating species, which in turn trigger the polymerization process of monomers. Typical instigators comprise organic peroxides, azo compounds, or redox mechanisms. In contrast, catalysts enable the polymerization process via an alternative mechanism, which may involve coordination with monomers or activation of particular functional groups. Catalysts commonly employed in latex polymerization include metal complexes, organometallic substances, and acid-base catalysts. The appropriate choice of initiators and catalysts plays a pivotal role in attaining the targeted polymerization speed, molecular weight, and characteristics of the resultant latex substrate (Touchard et al., 2004).

2.6.1. Initiators Used in Latex Polymerization Reactions

Initiators are essential in latex polymerization as they trigger the polymerization process by producing reactive species that create polymer chains from monomers. The selection of the initiator plays a pivotal role in determining the polymerization rate, molecular weight regulation, and the characteristics of the resultant latex material. This section will examine frequently utilized initiators in latex polymerization reactions.

2.6.1.1. Organic Peroxides

The utilization of organic peroxides as initiators within latex polymerization is prevalent owing to their capacity to undergo decomposition and produce free radicals. The polymerization process is initiated by the free radicals, which abstract hydrogen atoms from the monomers, thereby generating fresh reactive sites for the growth of the polymer chain. Cumene hydroperoxide, Benzoyl peroxide, and tert-butyl hydroperoxide are among the organic peroxides employed in the latex polymerization process. Tailoring the temperature at which decomposition occurs, and half-life for organic peroxides may be achieved to accommodate particular polymerization conditions (Hua et al., 2010).

2.6.1.2. Azo Compounds

Azo compounds represent a prevalent category of initiators that are frequently employed in the process of latex polymerization. The compounds above are characterized by the azo (-N=N-) functional moiety, which is susceptible to undergoing thermal or redox decomposition, leading to the generation of free radicals. Azo initiators are commonly activated by thermal means, necessitating heightened temperatures for the commencement of the polymerization process. Azo initiators, such as 2,2'-azobis(2-methylpropionitrile) (AMBN), azobisisobutyronitrile (AIBN), and 2,2'-azobis(2-amidinopropane) dihydrochloride (V-50), are commonly used in various chemical reactions. Azo initiators are recognized for their elevated efficacy and adept regulation of the polymerization procedure.

2.6.1.3. Redox Systems

Redox initiators are composed of an oxidizer and a reducer that undergo a reaction to produce free radicals. The initiators above exhibit significant utility in emulsion polymerization, wherein the redox reaction transpires at the surfaces of the latex particles. Redox initiators commonly comprise a blend of persulfates, such as ammonium persulfate, as the oxidizing agent and reducing agents, such as sodium bisulfite or sodium metabisulfite. The redox reaction induces the formation of unpaired electrons, known as free radicals, which serve as initiators for the polymerization process of monomers present in the latex system (Lamb et al., 2001).

2.6.1.4. Photoinitiators

Photoinitiators are initiators that trigger the polymerization process by being exposed to particular wavelengths of light. UV-curable latex systems are commonly utilized, wherein UV radiation exposure leads to swift polymerization. Photoinitiators can absorb ultraviolet radiation and subsequently undergo photolysis, producing free radicals or reactive species. These species serve as initiators for the process of polymerization. Benzoin ethyl ether, benzoin methyl ether, and benzoin isobutyl ether are the photoinitiators commonly employed in latex polymerization. Photoinitiators offer precise regulation of the polymerization procedure, enabling swift solidification and decreased energy demands.

2.6.1.5. Initiator Systems

In certain instances, initiator systems that amalgamate multiple initiators are utilized to attain specific polymerization traits. These systems can potentially enhance the regulation of polymerization rates, molecular weight distribution, and the polymerization of particular monomer types. Illustrative instances comprise the amalgamation of organic peroxides alongside azo initiators or the fusion of thermal catalysts with photoinitiators. Initiator systems offer versatility and facilitate customized polymerization conditions. The choice of a suitable initiator is contingent upon several factors, such as the intended polymerization rate, temperature, reactivity of the monomer, and stability prerequisites. The suitability of the initiator with the monomer system, its thermal endurance, and its decomposition characteristics are crucial factors to consider. The initiator concentration is crucial in attaining the intended regulation of polymerization kinetics and molecular weight (Joensson et al., 1994).

2.6.2. Redox Initiators and Their Mechanisms in Latex Chemistry

Redox initiators are crucial in latex chemistry due to their ability to produce free radicals by a combination of a reducing agent and an oxidizing agent. The initiators above are extensively employed in emulsion polymerization, wherein the redox reaction occurs at the surfaces of the latex particles. This section will delve into a more detailed examination of the processes and roles of redox initiators in latex chemistry.

2.6.2.1. Mechanism of Redox Initiation

The process of redox initiation comprises two distinct stages: initiation and propagation. The commencement phase commences with oxidizing the reducing agent through the oxidizing agent, producing unpaired and highly reactive free radicals. Persulfates, specifically ammonium persulfate, are conventionally utilized as oxidizing agents, whereas reducing agents such as sodium metabisulfite or bisulfite of sodium are frequently employed. The oxidation process of the reducing agent results in the generation of sulfate radicals, known for their high reactivity.

During the propagation stage, the sulfate radicals engage in a chemical reaction with monomers, extracting hydrogen atoms from the monomer molecules, thus instigating the polymerization process. This process generates novel unpaired electrons that persistently interact with monomers, expanding polymer chains. The latex particle creation and polymerization process is based on the reaction between sulfate radicals and monomers (Mohapatra & Nando, 2013).

2.6.2.2. Redox Initiation at the Particle Surfaces

The redox reaction occurs at the latex particles' surfaces during emulsion polymerization. The process involves the diffusion of the oxidizer and the reducer towards the exterior of the latex particles, generating free radicals through their reaction. The redox above reaction, which occurs in a confined region, facilitates the effective commencement of the polymerization mechanism, particularly on the surfaces of the particles.

2.6.2.3. Control over Polymerization Rate

Redox initiators possess the advantageous characteristic of enabling regulation of the polymerization rate. The kinetics of the redox reaction can be modulated by manipulating the concentrations of the oxidant and reductant species. By regulating the concentrations, it is possible to precisely adjust the polymerization rate to fulfill particular demands, such as attaining a targeted rate of latex particle expansion or obtaining a particular molecular weight of the resultant polymer (Jayachandran et al., 2002).

2.6.2.4. Effect on Molecular Weight Distribution

The molecular weight allocation of the polymer developed can also be affected by redox initiators. The beginning, propagation, and termination reaction rates influence the overall distribution of polymer chain lengths. By manipulating the concentrations of redox initiators and the reaction circumstances, it is feasible to customize the distribution of molecular weights, resulting in polymers that exhibit diverse properties and performance attributes.

2.6.2.5. Synergistic Effect

The interaction between the reducing and oxidizing agents within a redox initiator system produces a synergistic outcome that amplifies the initiation mechanism. The oxidizing agent facilitates the oxidation potential required for the production of radicals. In contrast, the reducing agent functions as an electron donor, which is expendable, to renew the oxidizing agent and maintain the redox reaction. The observed synergistic effect guarantees the uninterrupted starting and spreading of the polymerization process (Nenaah & Ahmed, 2011).

2.6.2.6. Application in Latex Chemistry

The widespread application of redox initiators in latex chemistry can be attributed to their water-based system compatibility and capacity to trigger polymerization at the surfaces of latex particles. These catalysts exhibit notable efficacy in polymerizing hydrophobic monomers such as styrene. The redox initiation mechanism facilitates the polymerization of monomers possessing diverse reactivity levels, thereby enabling the synthesis of a broad spectrum of latex materials that find application in many domains.

2.6.3. Catalysts and Their Role in Controlling Latex Polymerization

Catalysts are essential in latex polymerization as they facilitate the polymerization process and regulate diverse reaction features. The impact of catalysts on the polymerization rate, the molecular weight, and the characteristics of the resultant latex material is noteworthy. This section will delve deeper into the function of catalysts in regulating latex polymerization (Suddaby et al., 1996).

2.6.3.1. Activation of Functional Groups

Catalysts are frequently utilized to initiate particular functional groups in the monomers, thereby facilitating their polymerization. Acidic catalysts, namely phosphoric or sulfuric acid, can initiate hydroxyl groups in monomers such as acrylic acid or vinyl alcohol. The polymerization process is initiated by developing reactive species resulting from said activation. Catalysts influence the polymerization of particular monomers or combinations of monomers using targeted activation of specific functional groups.

2.6.3.2. Coordinating with Monomers

Certain catalysts operate through coordination with monomers, thereby enabling the process of polymerization. Metallic complexes, including organometallic compounds, can interact with monomers and facilitate their transformation into reactive species. Titanium-based catalysts, specifically titanium alkoxides, exhibit coordination with monomers with double bonds, augmenting their reactivity and expediting the polymerization process. Through coordination with monomers, catalysts facilitate effective initiation and regulation of the polymerization process (Held & Mecking, 2000).

2.6.3.3. Controlling Polymerization Rate

Catalysts are essential in regulating the rate of polymerization in latex systems. Polymers' overall polymerization rate is determined by their influence on the initiation, growth, and termination reactions. It is feasible to attain the targeted polymerization rate for a given latex formulation through meticulous catalyst selection and concentration adjustments. Regulating the polymerization rate is crucial in attaining homogeneous particle expansion, managing the molecular weight, and securing the intended characteristics of the resultant latex substance.

2.6.3.4. Molecular Weight Control

The molecular weight of the polymer produced during latex polymerization can be affected by catalysts. The extent of chain propagation and termination can be influenced by factors such as catalyst selection, concentration, and reaction conditions. By manipulating these parameters, it is feasible to regulate the polymer chain length, consequently governing the molecular weight of the latex substance. The regulation of molecular weight is a crucial factor in attaining the ultimate latex commodity's intended attributes and functional features (Smulders & Monteiro, 2004).

2.6.3.5. Polymerization Mechanism

Catalysts can influence the polymerization process in latex systems. Diverse catalysts can facilitate distinct polymerization mechanisms, including anionic polymerization, radical polymerization, and cationic polymerization. The mechanism of polymerization is contingent upon the selection of the catalyst. Every polymerization mechanism possesses its own set of benefits and has the potential to yield latex materials with distinct properties. The judicious choice of a suitable catalyst enables regulation of the polymerization mechanism, thereby influencing the characteristics of the resultant latex material.

2.6.3.6. Thermal or Photochemical Activation

The catalysts can be activated via thermal or photo-chemical methods, contingent upon the particular catalyst employed. The thermal activation process entails subjecting the latex system to a particular temperature, which triggers the catalyst's activation.

In contrast, photochemical activation entails the utilization of particular light wavelengths to initiate the catalyst. Using photochemically activated catalysts is especially advantageous in UV-curable latex systems due to their ability to initiate polymerization upon exposure to UV radiation promptly. The selection between photochemical or thermal activation depends on the specific latex formulation, processing specifications, and intended curative characteristics (Corrigan et al., 2020).

2.7. LATEX PARTICLE SIZE ANALYSIS

Analyzing particle size in latex is a critical component in the characterization of latex dispersions. The rheological behavior, stability, and efficiency of latex materials in diverse applications are directly impacted by particle size. The examination and comprehension of the distribution of particle sizes enable the enhancement of formulation and manufacturing parameters to attain the intended characteristics and functionality. Several methodologies are available for analyzing latex particle size, comprising dynamic light scattering (DLS), laser diffraction, microscopy, and sedimentation techniques (e.g., centrifugal sedimentation). The techniques above furnish significant insights into the distribution of particle size, mean particle size, and the potential occurrence of agglomeration or coalescence during the production or storage of latex. The ability to conduct particle size analysis accurately and precisely is crucial for researchers and manufacturers to customize latex formulations and procedures to achieve optimal performance and meet specific application requirements (Duracher et al., 1998) (Figure 2.10).

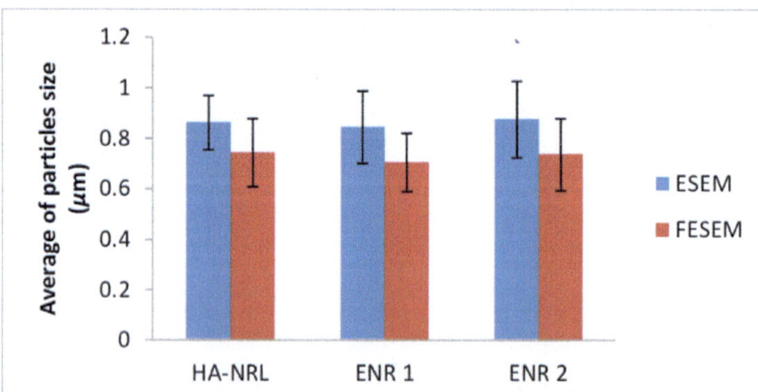

Figure 2.10. *Example of average latex particles size for different latexes based at latex (ESEM) and dry state (FESEM).*

Source: Norhanifah mohd yazid, Creative Commons License.

2.7.1. Techniques for Measuring Particle Size in Latex Dispersions

The precise determination of particle size is crucial in comprehending latex dispersions' characteristics and efficacy. The particle size distribution directly influences the rheological behavior, stability, optical characteristics, and functionality of latex materials in diverse applications. Various methodologies exist for determining the particle's size in latex dispersions, each possessing distinct merits and demerits. This section will examine frequently employed methods for measuring particle size in latex dispersions.

2.7.1.1. Dynamic Light Scattering (DLS)

The technique of Dynamic Light Scattering, commonly referred to as Photon Correlation Spectroscopy, is extensively employed in determining the size distribution of particles in latex dispersions. DLS is a technique that quantifies the variations in the intensity of scattered light resulting from the random movement of particles, also known as Brownian motion. The distribution of particle sizes and average particle size can be determined by analyzing intensity fluctuations. DLS is a highly appropriate technique for determining minute particle sizes, typically in the range of a few nanometers to micrometers. The method is characterized by its expeditious outcomes and applicability to undiluted latex dispersions, rendering it a prevalent option for customary particle size evaluation (Gugliotta et al., 2000) (Figure 2.11).

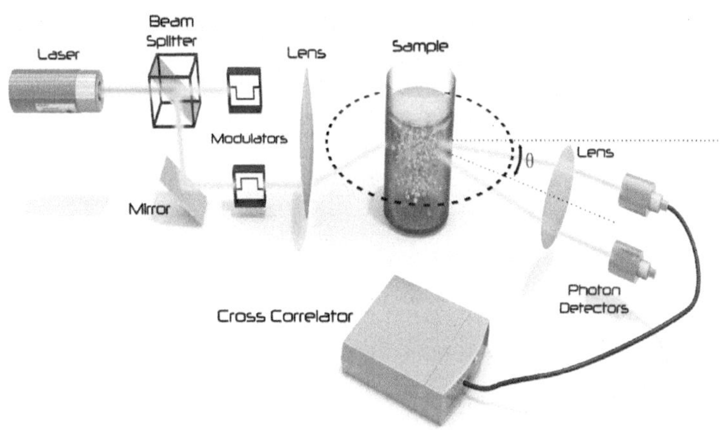

Figure 2.11. *Schematic of the dynamic light scattering (DLS) technique.*

Source: Nicole Meulendijks, Creative Commons License.

2.7.1.2. Laser Diffraction

The technique of Laser Diffraction is frequently employed to analyze particle size. It applies to a broad spectrum of particle sizes ranging from sub-micron to a few hundred micrometers. The method employed involves the assessment of the dispersion's particle scattering pattern generated upon the interaction of a laser beam with said particles. Subsequently, the scattering pattern is subjected to analysis to ascertain the distribution of particle sizes. Laser diffraction presents a prompt and dependable means of obtaining measurements while also furnishing insights into the mean particle size and the distribution of particle sizes. It is noteworthy that laser diffraction methodology operates under the assumption of spherical particle shapes, which could potentially result in an underestimation of particle size for non-spherical or irregularly shaped particles.

2.7.1.3. Microscopy

Microscopy methodologies, such as Transmission Electron Microscopy (TEM) and SEM, facilitate the direct observation and quantification of latex particles. Utilizing these methodologies facilitates the acquisition of high-resolution images, thereby enabling the determination of the sizes of individual particles, characterization of their shapes, and evaluation of their morphology. The SEM technique is advantageous in examining larger-sized particles. In contrast, the TEM method is more appropriate for analyzing smaller particles within the nanometer scale. Microscopy techniques offer significant qualitative and quantitative insights into particles' dimensions, morphology, and surface properties. Nevertheless, statistical evaluation and large-scale particle size examinations may encounter obstacles due to sample preparation and imaging limitations (Dagata et al., 2016).

2.7.1.4. Sedimentation Methods

The sedimentation process entails determining the velocity at which latex particles settle when subjected to the effects of gravitation or centrifugal force. Using sedimentation methods, namely gravity sedimentation and centrifugal sedimentation, is feasible for determining particle size distribution and density of particles in latex dispersions. The sedimentation behavior can be utilized to determine the particle size as the settling rate exhibits an inverse relationship with the particle size. Sedimentation techniques are especially well-suited for analyzing particle sizes that are relatively large and can furnish valuable insights into both the distribution of particle sizes and the stability of settling. Nevertheless, utilizing these methodologies necessitates meticulous sample preparation, contemplation of particle density, and regulation of experimental variables to attain precise and dependable outcomes.

2.7.1.5. Nanoparticle Tracking Analysis (NTA)

The NTA method is a hybrid approach that integrates microscopy and Brownian motion assessment to quantify the dimensions and abundance of nano-particles in latex dispersions. The method employed involves the utilization of video microscopy to monitor the movement of individual particles and subsequently analyze their Brownian motion to determine their respective sizes. The utilization of NTA is deemed appropriate for diminutive particle dimensions, typically encompassing a range of a few nanometers to a few hundred nanometers in size. This offers insights into the distribution of particle sizes, concentration levels, and aggregation patterns. Nevertheless, the NTA technique may exhibit constraints regarding precision and detail when dealing with bigger particle dimensions and elevated particle densities (Boyd et al., 2011) (Figure 2.12).

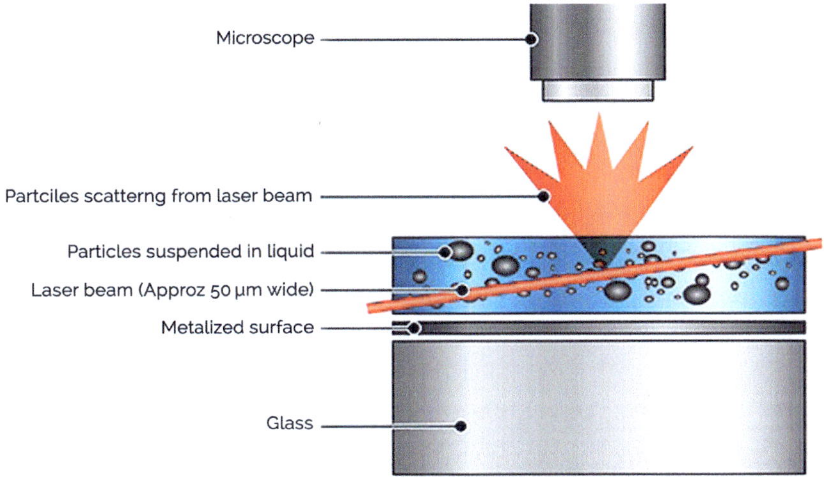

Figure 2.12. *Nanoparticle tracking analysis (NTA) measurement technique.*

Source: Azonano, Creative Commons License.

2.7.1.6. Electrical Sensing Zone (Coulter Counter)

The Electrical Sensing Zone, or Coulter Counter, is a method utilized to quantify alterations in electrical resistance that occur when particles traverse a minute aperture. The particle size distribution can be inferred from the information provided by the resistance change in terms of its magnitude and duration. The Electrical Sensing Zone method is well-suited for analyzing latex dispersions that exhibit a wide range of particle sizes, ranging from sub-micron up to a few hundred micrometers. The measurements provided by the method are highly accurate and precise and are not influenced by either the refractive index or the shape of the particles. Nevertheless, implementing this methodology necessitates appropriate sample dilution and meticulous regulation of experimental variables to guarantee precise particle size analysis.

2.7.2. Importance of Particle Size Analysis in Latex Characterization

The evaluation of latex materials is significantly reliant on particle size analysis. The article presents significant insights into the dispersion, mean dimensions, and structure of latex particles, which directly influence their endurance, flow properties, visual characteristics, and utility in diverse domains. The comprehension and regulation of particle size are crucial for optimizing latex composition, processing parameters, and attaining intended performance attributes. The significance of the analysis of particle sizes in the characterization of latex shall be examined in greater depth.

2.7.2.1. Stability and Shelf Life

The evaluation of particle size is of utmost importance in determining the stability and duration of storage of latex dispersions. The stability of latex materials is influenced by the distribution of particle sizes, wherein the presence of smaller particles promotes suspension. In contrast, larger particles are prone to settling or agglomeration with time. Through monitoring particle size distribution, alterations in stability can be identified, thereby facilitating modifications in formulation or conditions of storage to augment stability and extend the shelf life. The particle size analysis offers valuable insights into various factors, including flocculation, coagulation, and Ostwald ripening, which can significantly affect the stability of latex dispersions (Wen et al., 2002).

2.7.2.2. Rheological Behavior

The size of particles directly impacts the rheological properties of latex materials. The properties of latex dispersions, specifically the viscosity shear thinning shear recovery, are impacted by the distribution of particle sizes. Reduced particle size results in augmented surface area and heightened surface charge, thereby contributing to elevated viscosity and improved shear resistance. Conversely, particles of greater size are known to decrease viscosity and facilitate fluidity. The comprehension and regulation of the rheological properties of latex dispersions can be facilitated through precise particle size measurement, thereby allowing for the enhancement of formulation and manufacturing parameters.

2.7.2.3. Film Formation and Coating Properties

The particle size analysis is of utmost importance in comprehending the film formation process and enhancing the properties of coatings in coatings based on latex. The particle size distribution influences the packing density and arrangement of particles in the dried film, thereby affecting different characteristics such as smoothness, film thickness, adhesion, and outer layer appearance. Through analyzing and regulating the particle size distribution, it is possible to customize the coating characteristics to fulfill particular demands, such as attaining a consistent and flawless film with the intended thickness and surface texture (Kablitz & Urbanetz, 2007).

2.7.2.4. Optical Properties

Analyzing particle size is crucial in comprehending and managing the optical characteristics of latex materials. The properties of latex-based coatings, inks, and films, such as transparency, opacity, gloss, and color, are influenced by the particles' size. The size of particles has an impact on the optical properties of a material. Smaller particles enhance transparency and gloss, whereas larger particles decrease gloss and increase opacity. The optimization of optical features of latex materials for specific uses, such as attaining the desired degree of transparency, intensity of color, or light scattering behavior, can be achieved through precise measurement of particle size.

2.7.2.5. Processability and Stability of Emulsions

The analysis of particle size plays a crucial role in assessing the processing and stability of latex formulations during their formulation and processing. The size of particles influences the viscosity, stability, and handling of emulsions. The reduction in particle size results in decreased viscosity and improved stability, facilitating the blending, pumping, and application processes. Adjustments to formulation and processing surroundings can be made by monitoring particle size to attain steady emulsions with optimal processability. This can effectively mitigate problems such as obstruction, uneven covering, or poor film formation (Wu et al., 2011).

2.7.2.6. Control of Functional Properties

The analysis of particle size is a crucial aspect in the management of the operational characteristics of latex materials. The properties of elasticity, mechanical strength, thermal stability, barrier characteristics, and surface reactivity are influenced by the size of the particles. The precise measurement and regulation of particle size are crucial in attaining the intended functional characteristics of latex materials. The particle size directly influences the final product's flexibility, adhesive strength, and elongation characteristics in latex-based adhesives or elastomers.

2.7.2.7. Quality Control and Process Optimization

The particle size analysis is crucial for optimizing processes and quality control in latex production. Frequent particle size monitoring enables the detection of deviations from predetermined specifications, facilitating the prompt implementation of corrective measures.

The particle size analysis facilitates process optimization by comprehending process variables' impact on particle size. This allows for modifying formulation or manufacturing conditions to attain the intended particle size distribution. The implementation of this measure guarantees uniformity in the quality of the product, its replicability, and the efficacy of latex materials (Buron et al., 2004).

2.8. MOLECULAR WEIGHT DETERMINATION

Precise molecular weight calculation is a crucial aspect of comprehending and managing the characteristics of latex polymers. Various methodologies exist for ascertaining molecular weight, including size exclusion chromatography (SEC), viscosity measurements, light scattering techniques, and gel permeation chromatography (GPC). The techniques above offer significant insights into the distribution of polymer chain length and the mean molecular weight. Determining molecular weight facilitates formulation optimization, the process, and the efficiency of latex materials for particular applications by researchers and manufacturers (Goetz et al., 2004).

2.8.1. Methods for Determining Molecular Weight

Precise molecular weight calculation is of utmost importance in latex chemistry as it offers significant insights into the properties, structure, and performance of the polymer. There exist various methodologies to ascertain the molecular weight of latex polymers. This section will examine frequently employed techniques for determining molecular weight in latex chemistry.

Remember

When characterizing latex polymers, it is important to determine their molecular weight using techniques like gel permeation chromatography (GPC) or size exclusion chromatography (SEC).

2.8.1.1. Gel Permeation Chromatography (GPC)

Determining the distribution of molecular weight of latex polymers is commonly achieved through utilizing GPC or SEC, which are highly prevalent techniques. The process involves the separation of polymers by size as they traverse a porous stationary stage. The fundamental concept underpinning this methodology is that the swifter elution of more substantial polymer molecules compared to their smaller counterparts facilitates the ascertainment of the distribution of molecular weights. The calibration of GPC/SEC necessitates the utilization of standard polymers with established molecular weights to create a calibration curve. This methodology yields data on the mean molecular weight, distribution, and polydispersity index (PDI) of the latex polymer (Timpa, 1991).

2.8.1.2. Viscometry

Viscometry is a commonly employed and uncomplicated method for determining the mean molecular weight of latex polymers. The present methodology involves the determination of the viscosity of a sparse polymer solution, which is subsequently associated with its molecular weight through the utilization of a Mark-Houwink equation. The determination of the viscosity of a polymer solution is commonly accomplished through the utilization of a viscometer. Meanwhile, the molecular weight of the said polymer is ascertained by comparing its viscosity with that of typical polymer solutions whose molecular weights are already known. Viscometry is a valuable technique for determining the molecular weight of soluble polymers in solution, providing a relative measure of their properties.

2.8.1.3. Light Scattering Techniques

Utilizing light scattering methodologies, including static light scattering and dynamic light scattering, is a valuable means of obtaining information regarding the molecular weight and shape of latex polymers. Static Light Scattering (SLS) enables the evaluation of the mean molecular weight of a polymer by measuring the scattered light intensity concerning the scattering angle. Dynamic Light Scattering (DLS) is a technique that analyzes scattered light intensity fluctuations resulting from the Brownian motion of particles. This method offers valuable insights into the size and distribution of latex polymers. Light scattering methodologies are non-invasive and can be executed on undiluted latex dispersions (Wedlock et al., 1986).

2.8.1.4. Ultracentrifugation

The ultracentrifugation process involves the separation of particles through their settling rate in a centrifugal field. The molecular weight of latex polymers can be approximated by using this technique to evaluate the sedimentation coefficient of particles. The relationship between the sedimentation coefficient and particle size and shape can be utilized to establish a correlation with the molecular weight. Ultracentrifugation is a technique that yields insights into the latex polymer's hydrodynamic diameter and molecular weight.

2.8.1.5. Mass Spectrometry

The utilization of mass spectrometry as an analytical method is highly effective in the determination of both the molecular weight and the structural properties of latex polymers. The process entails the ionization of polymer molecules, which is subsequently followed by determining their mass-to-charge ratio. Mass spectrometry is a highly informative analytical technique that can yield accurate and comprehensive data regarding the molecular weight, end-group performance, and sequence distribution of latex polymers. The characterization of complex polymer structures and copolymers is notably advantageous (Covey et al., 1988).

2.8.1.6. Nuclear Magnetic Resonance (NMR) Spectroscopy

The technique of NMR spectroscopy is widely employed to ascertain the molecular weight and makeup of latex polymers. NMR spectroscopy is a valuable analytical technique that can offer insights into the connectivity and structure of monomers within a polymer chain. This information can be utilized to ascertain the polymer's molecular weight and branched features. Furthermore, diffusion-ordered spectroscopy (DOSY) can offer valuable perspectives on the latex polymer's dimensions and molecular weight dispersion. The selection of a technique for ascertaining the molecular weight in the domain of latex chemistry is contingent upon various factors, such as the characteristics of the polymer, the availability of the sample, the accessibility of equipment, and the intended degree of precision. Integrating various methodologies can yield a more exhaustive comprehension of the latex polymer's size distribution, molecular weight, and structure. The precise determination of molecular weight facilitates the optimization of formulation, process, and performance of latex materials for particular applications in various industries, including covers, adhesives and related products, textiles, and biomedical materials, by researchers and manufacturers (Daykin et al., 2002).

2.8.2. Significance of Molecular Weight in Latex Polymer Properties

Their molecular weight significantly influences the determination of latex polymers' properties and functioning characteristics. The polymer is impacted across multiple dimensions, encompassing its physical characteristics, rheological performance, mechanical potency, manufacturability, and interfacial dynamics with other substances. This section will delve deeper into the importance of molecular weight concerning the properties of latex polymers.

2.8.2.1. Physical Properties

The physical properties of latex polymers are directly influenced by their molecular weight. Polymers with a greater molecular weight typically exhibit heightened density, viscosity, and molecular dimensions. The film formation, flow behavior, and mechanical properties associated with latex material are influenced by various factors. This phenomenon illustrates the enhanced durability, elasticity, and impact resistance observed in polymers with higher molecular weights, which can be attributed to heightened chain entanglement and forces between molecules (Martinsen et al., 1991).

2.8.2.2. Rheological Behavior

The rheological properties of latex polymers are significantly impacted by their molecular weight. Elevated molecular weights of polymer chains are associated with augmented viscosity and enhanced shear resistance. This phenomenon impacts the flow characteristics and processing of the latex material during the covering, casting, or extrusion processes. The capacity to regulate the molecular weight confers the possibility of customizing the rheological characteristics to fulfill particular application demands, such as attaining the intended coating thickness or managing the flow behavior throughout the application process.

2.8.2.3. Solubility and Swelling

The molecular weight of latex polymers influences their solubility and swelling behavior. Polymers with lower molecular weights possess superior solubility and quicker dissolution rates in solvents. In contrast, those with higher molecular weights may necessitate more effort for dissolution. The molecular weight impacts the swelling behavior, including the uptake of water or solvent. The characteristics above hold significant importance in coatings, bonds, and drug delivery systems applications, wherein the solubility and expanding behavior play a crucial role in determining the film formation, adherence, and release properties (Painter et al., 1990).

2.8.2.4. Mechanical Properties

The mechanical characteristics of latex polymers are directly impacted by their molecular weight. Higher molecular weight polymers' elevated tensile strength, modulus, and toughness can be attributed to the augmented chain entanglement and interactions between molecules.

The enhanced mechanical properties of these materials render them appropriate for deployment in scenarios that necessitate robustness or resilience, such as in the domains of coatings, glues, and elastomers. On the other hand, polymers with reduced molecular weight could potentially provide enhanced flexibility and increased elongation at the fracture point.

2.8.2.5. Processing and Film Formation

The processing and film formation features of latex polymers are impacted by their molecular weight. The film formation process for polymers with higher molecular weights typically necessitates greater shear forces or extended drying periods.

Those above can affect the process variables and parameters, including but not limited to coating thickness, drying duration, and curing temperature. The regulation of molecular weight facilitates the optimization of processing parameters, thereby enabling the attainment of needed film formation, adherence, and surface properties (Chi et al., 2021).

2.8.2.6. Polymer Chain Entanglement

The molecular weight influences the level of polymer chain entanglement within latex polymers. Polymers with elevated molecular weight exhibit elongated chains, resulting in augmented entanglement and inter-penetration regarding polymer chains. The interconnection between the constituent particles of the latex material is responsible for its viscoelastic properties, as well as its melt strength and elasticity. The entanglement of polymer chains is known to confer superior resistance to movement and deformation, thereby augmenting the durability and mechanical characteristics of the latex polymer.

2.8.2.7. Copolymerization and Blend Compatibility

The molecular weight of every constituent can influence the suitability and phase behavior of copolymers or polymer mixtures. The variation can influence the suitability and separation of phase tendencies of a blend of polymers in their molecular weights. The manipulation of the molecular weight proportion between the constituent elements allows for adjusting the characteristics of the composite material, including its shape, thermal endurance, and mechanical proficiency (Chiu & Smith, 1984).

2.8.3. Factors Affecting Molecular Weight in Latex Polymerization

The polymerization process of latex polymers is influenced by multiple factors that impact their molecular weight. Comprehending these variables is imperative in regulating and customizing the molecular weight to attain the intended attributes and operational features. This section will examine the fundamental factors that impact molecular weight during latex polymerization.

2.8.3.1. Monomer Reactivity

The impact of monomer reactivity on the molecular weight is a critical consideration. Monomers exhibiting greater reactivity typically undergo polymerization at a faster rate, resulting in the formation of shorter chains of polymers and a reduction in molecular weight. On the contrary, monomers that exhibit lower reactivity may lead to the formation of lengthier polymer chains and an elevated molecular weight. The selection of monomers and their respective reactivity ratios are crucial factors in determining the mean molecular weight of the resultant latex polymer (Schoonbrood & Asua, 1997).

2.8.3.2. Initiator Concentration

The concentration of initiators present in the polymerization system influences the polymerization rate and molecular weight. Elevated concentrations of initiators typically result in accelerated polymerization kinetics, leading to reduced molecular weight and shorter polymer chains. In contrast, reduced initiator concentrations decelerate the polymerization rate, thereby facilitating the formation of extended polymer chains and elevated molecular weights. Regulating the molecular weight throughout latex polymerization can be achieved by regulating initiator concentration.

2.8.3.3. Monomer Concentration

The relationship between the monomer concentration and the resulting molecular weight is direct. Elevated concentrations of monomers frequently lead to decreased molecular weight and shorter polymer chains due to heightened occurrences of chain termination. In contrast, reduced concentrations of monomers facilitate the formation of lengthier polymer chains and elevated molecular weights. Precise regulation of monomer concentration is essential for achieving the targeted molecular weight and promoting the homogeneous expansion of latex particles (Maxwell et al., 1992).

2.8.3.4. Reaction Temperature

The reaction's temperature influences the total molecular weight of the polymer produced during the latex polymerization process. Elevated temperatures typically induce accelerated polymerization kinetics, reducing molecular weight and shorter polymer chains. The reduction of temperature results in a deceleration of the polymerization rate, leading to the formation of extended polymer chains and an increase in molecular weight. Optimization of the reaction temperature is crucial in achieving the intended molecular weight of the polymer, considering the reactivity and stability of both the monomers and initiators.

2.8.3.5. Chain Transfer Agents

Utilizing chain transfer agents is a viable approach to regulating molecular weight. The agents above function as molecular weight regulators by transferring a polymer chain in its nascent stage from its active location to the chain-transferring agent. The process above is a proficient means of concluding the polymer chain, thereby reducing molecular weight and shorter chains. Thiols, mercaptans, and halogenated compounds are frequently employed as chain transfer agents. The regulation of the molecular weight of a latex polymer can be achieved by adjusting the amount present in the chain transfer agent.

2.8.3.6. Polymerization Time

The molecular weight is also influenced by the length of time of the polymerization process. Increased duration of polymerization facilitates the greater conversion of monomers and chain elongation, leading to the formation of extended polymer chains and elevated molecular weight. Reduced polymerization durations constrain the elongation of chains, resulting in decreased polymer chain length and diminished molecular weight. Precise regulation of the polymerization duration is imperative to attain the targeted molecular weight, concurrently guaranteeing satisfactory conversion and particle expansion of the monomer (Charleux et al., 2010).

2.8.3.7. Stirring Rate

The stirring rate employed during latex polymerization influences the polymer's molecular weight. Elevated stirring rates have the potential to augment the frequency of collisions between nascent chains of polymers and monomers, thereby expediting chain propagation and producing polymer chains of diminished molecular weight. Reducing the stirring rates results in a decrease in collision frequency, which facilitates the formation of longer chains of polymers and consequently leads to an increase in molecular weight. Optimization of the stirring rate is necessary to attain the intended molecular weight while considering the constraints imposed by the transfer of heat and mass transfer (Hayashi et al., 1989).

Activity 2.1

Suppose you are a researcher working on latex chemistry. You are studying the effect of different crosslinking agents on the properties of latex materials. Design an experiment to compare the properties of two latex samples, one crosslinked with a conventional crosslinking agent and the other crosslinked with a newly developed crosslinking agent. How would you analyze and compare the properties of these latex samples?

2.9. SUMMARY

The chapter titled "Fundamentals of Latex Chemistry" presents a thorough exposition of the fundamental facets of latex chemistry, with a particular emphasis on the comprehension of latex polymerization, particle generation, and the influence of diverse factors on latex characteristics. The introductory section of the chapter emphasizes the significance of comprehending the chemistry of latex in the advancement and enhancement of latex-based materials. The chapter delves into various subjects, including the chemical makeup of latex particles and the function of monomers in latex polymerization. The present study investigates the influence of crosslinking agents on the properties of latex and scrutinizes the diverse polymerization mechanisms, namely anionic, radical, cationic, and copolymerization, in latex systems. This discourse delves into the configuration and structure of latex particles, encompassing the mechanisms of nucleation and coagulation alongside the methodologies of emulsion polymerization. The chapter analyzes the shape and form of latex particles, focusing on their impact on the properties and applications of latex. This study delves into the examination of surfactants and stabilizers in latex chemistry, with a specific emphasis on their respective roles in the stabilization and dispersion of latex.

Furthermore, the chapter delves into the significance of initiators as well as catalysts in the process of latex polymerization, examining their influence on both the polymerization mechanism and the resultant properties of the latex. This article delves into molecular weight estimation techniques and their relevance in the characterization of latex. It offers helpful information into the calculation of molecular weight and its impact on the properties of latex. The final section of the chapter provides an analysis of the significance of particle size assessment in the characterization of latex, emphasizing its influence on the rheology, stability, film formation, and working attributes of latex substances. This study investigates the various factors that affect particle size distribution in latex systems, including concentration, monomer reactivity, temperature, mixing, stabilizers, and polymerization time. In general, the publication "Fundamentals of Latex Chemistry" furnishes an all-encompassing comprehension of the fundamental principles and variables implicated in latex chemistry. The chapter provides readers with comprehensive information regarding the chemical makeup, the polymerization processes, particle development, and characterization techniques that are essential for developing and optimizing latex materials suitable for a diverse array of applications.

REVIEW QUESTIONS

1. What is the chemical composition of latex particles?
2. Explain the role of monomers in latex polymerization.
3. How do crosslinking agents impact the properties of latex?
4. Discuss the different polymerization mechanisms in latex, including radical, anionic, cationic, and copolymerization.
5. Describe the processes of nucleation and coagulation in latex particle formation.
6. What are the different emulsion polymerization techniques used in latex production?
7. Explain the role of surfactants and stabilizers in latex chemistry.
8. What initiators are used in latex polymerization reactions?
9. Discuss the techniques for measuring particle size in latex dispersions and their significance.
10. How is molecular weight determined in latex polymers, and what factors affect it in polymerization?

REFERENCES

1. Abele, S., Zicmanis, A., Graillat, C., & Guyot, A., (1999). Cationic and zwitterionic polymerizable surfactants quaternary ammonium dialkyl maleates. 2. Emulsion polymerization of styrene and butyl acrylate. *Langmuir, 15*(4), 1045–1051.
2. Aoshima, S., & Kanaoka, S., (2009). A renaissance in living cationic polymerization. *Chemical Reviews, 109*(11), 5245–5287.
3. Aramendia, E., Mallégol, J., Jeynes, C., Barandiaran, M. J., Keddie, J. L., & Asua, J. M., (2003). Distribution of surfactants near acrylic latex film surfaces: A comparison of conventional and reactive surfactants (surfmers). *Langmuir, 19*(8), 3212–3221.
4. Ata, S., Davis, E. S., Dupin, D., Armes, S. P., & Wanless, E. J., (2010). Direct observation of pH-induced coalescence of latex-stabilized bubbles using high-speed video imaging. *Langmuir, 26*(11), 7865–7874.
5. Bilgin, S., Tomovska, R., & Asua, J. M., (2016). Fundamentals of chemical incorporation of ionic monomers onto polymer colloids: Paving the way for surfactant-free waterborne dispersions. *RSC Advances, 6*(68), 63754–63760.
6. Boyd, R. D., Pichaimuthu, S. K., & Cuenat, A., (2011). New approach to inter-technique comparisons for nanoparticle size measurements; using atomic force microscopy, nanoparticle tracking analysis and dynamic light scattering. *Colloids and Surfaces A: Physicochemical and Engineering Aspects, 387*(1–3), 35–42.
7. Braunecker, W. A., & Matyjaszewski, K., (2007). Controlled/living radical polymerization: Features, developments, and perspectives. *Progress in Polymer Science, 32*(1), 93–146.
8. Buron, H., Mengual, O., Meunier, G., Cayre, I., & Snabre, P., (2004). Optical characterization of concentrated dispersions: Applications to laboratory analyses and on - line process monitoring and control. *Polymer International, 53*(9), 1205–1209.
9. Cal, J. C. D. L., Urzay, R., Zamora, A., Forcada, J., & Asúa, J. M., (1990). Simulation of the latex particle morphology. *Journal of Polymer Science Part A: Polymer Chemistry, 28*(5), 1011–1031.
10. Castellanos, S. G., Fernández-Escamilla, V. V. A., Corona-Rivera, M. Á., González-Iñiguez, K. J., Barrera, A., Moscoso-Sánchez, F. J., & Aguilar, J., (2023). coagulative nucleation in the copolymerization of methyl methacrylate–butyl acrylate under monomer-starved conditions. *Polymers, 15*(7), 1628.

11. Ceska, G. W., (1974). Carboxyl - stabilized emulsion polymers. *Journal of Applied Polymer Science, 18*(8), 2493–2499.

12. Chai, X. S., Hou, Q. X., & Schork, F. J., (2004). Determination of residual monomer in polymer latex by full evaporation headspace gas chromatography. *Journal of Chromatography A, 1040*(2), 163–167.

13. Charleux, B., D'Agosto, F., & Delaittre, G., (2010). Preparation of hybrid latex particles and core–shell particles through the use of controlled radical polymerization techniques in aqueous media. *Hybrid Latex Particles: Preparation with (Mini) Emulsion Polymerization*, 125–183.

14. Chern, C. S., (2006). Emulsion polymerization mechanisms and kinetics. *Progress in Polymer Science, 31*(5), 443–486.

15. Chi, H. L., Yeh, A. I., Pan, M. H., & Chen, S. H., (2021). Physicochemical properties and film formation of the chitin hydrocolloid fabricated by a novel green process. *Journal of Applied Polymer Science, 138*(31), 50762.

16. Chiu, S. C., & Smith, T. G., (1984). Compatibility of poly (ϵ - caprolactone) (PCL) and poly (styrene - co - acrylonitrile)(SAN) blends. II. The influence of the AN content in SAN copolymer upon blend compatibility. *Journal of Applied Polymer Science, 29*(5), 1797–1814.

17. Cho, I., & Lee, K. W., (1985). Morphology of latex particles formed by poly (methyl methacrylate) - seeded emulsion polymerization of styrene. *Journal of Applied Polymer Science, 30*(5), 1903–1926.

18. Corrigan, N., Jung, K., Moad, G., Hawker, C. J., Matyjaszewski, K., & Boyer, C., (2020). Reversible-deactivation radical polymerization (Controlled/living radical polymerization): From discovery to materials design and applications. *Progress in Polymer Science, 111*, 101311.

19. Covey, T. R., Bonner, R. F., Shushan, B. I., Henion, J., & Boyd, R. K., (1988). The determination of protein, oligonucleotide and peptide molecular weights by ion - spray mass spectrometry. *Rapid Communications in Mass Spectrometry, 2*(11), 249–256.

20. Dagata, J., Kavuri, P., Vladár, A., Ehara, K., Farkas, N., Wu, C. L., & Itoh, H., (2016). Method for measuring the diameter of polystyrene latex reference spheres by atomic force microscopy. *Natl. Inst. Stand. Technol. Spec. Publ., 41, 260*, 185.

21. David, G., Boyer, C., Tonnar, J., Ameduri, B., Lacroix-Desmazes, P., & Boutevin, B., (2006). Use of iodocompounds in radical polymerization. *Chemical Reviews, 106*(9), 3936–3962.

22. Daykin, C. A., Foxall, P. J., Connor, S. C., Lindon, J. C., & Nicholson, J. K., (2002). The comparison of plasma deproteinization methods for the

detection of low-molecular-weight metabolites by 1H nuclear magnetic resonance spectroscopy. *Analytical Biochemistry, 304*(2), 220–230.

23. Deng, Y., Xiao, H., & Pelton, R., (1996). Temperature-sensitive flocculants based on poly (N-isopropylacrylamide-co-diallyldimethylammonium chloride). *Journal of Colloid and Interface Science*, *179*(1), 188–193.
24. Dimonie, V., El - Aasser, M. S., Klein, A., & Vanderhoff, J. W., (1984). Core - shell emulsion copolymerization of styrene and acrylonitrile on polystyrene seed particles. *Journal of Polymer Science: Polymer Chemistry Edition*, *22*(9), 2197–2215.
25. Dingenouts, N., Norhausen, C., & Ballauff, M., (1998). Observation of the volume transition in thermosensitive core– shell latex particles by small-angle X-ray scattering. *Macromolecules*, *31*(25), 8912–8917.
26. Duracher, D., Sauzedde, F., Elaissari, A., Perrin, A., & Pichot, C., (1998). Cationic amino-containing N-isopropyl-acrylamide–styrene copolymer latex particles: 1-Particle size and morphology vs. polymerization process. *Colloid and Polymer Science*, *276*, 219–231.
27. Dworak, A., Walach, W., & Trzebicka, B., (1995). Cationic polymerization of glycidol. Polymer structure and polymerization mechanism. *Macromolecular Chemistry and Physics*, *196*(6), 1963–1970.
28. Fang, C., & Lin, Z., (2015). Effect of propyleneimine external cross-linker on the properties of acrylate latex pressure sensitive adhesives. *International Journal of Adhesion and Adhesives*, *61*, 1–7.
29. Feeney, P. J., Gilbert, R. G., & Napper, D. H., (1985). Periodic nucleation processes in emulsion polymerization systems. *Journal of Colloid and Interface Science*, *107*(1), 159–173.
30. Feeney, P. J., Napper, D. H., & Gilbert, R. G., (1987). Surfactant-free emulsion polymerizations: Predictions of the coagulative nucleation theory. *Macromolecules*, *20*(11), 2922–2930.
31. Fuchs, A., & Killmann, E., (2001). Adsorption of polyelectrolytes on colloidal latex particles, electrostatic interactions and stability behavior. *Colloid and Polymer Science*, *279*, 53–60.
32. Fujii, S., Suzaki, M., Armes, S. P., Dupin, D., Hamasaki, S., Aono, K., & Nakamura, Y., (2011). Liquid marbles prepared from pH-responsive sterically stabilized latex particles. *Langmuir*, *27*(13), 8067–8074.
33. Gabriel, V. A., Champagne, P., Cunningham, M. F., & Dubé, M. A., (2022). In - situ addition of carboxylated cellulose nanocrystals in seeded semi - batch emulsion polymerization. *The Canadian Journal of Chemical Engineering*, *100*(4), 767–779.

34. Gan, L. H., & Ng, S. C., (1986). Kinetic studies of the performic acid epoxidation of natural rubber latex stabilized by cationic surfactant. *European Polymer Journal*, *22*(7), 573–576.

35. Garcia-Garcia, S., Wold, S., & Jonsson, M., (2009). Effects of temperature on the stability of colloidal montmorillonite particles at different pH and ionic strength. *Applied Clay Science*, *43*(1), 21–26.

36. Goetz, H., Kuschel, M., Wulff, T., Sauber, C., Miller, C., Fisher, S., & Woodward, C., (2004). Comparison of selected analytical techniques for protein sizing, quantitation, and molecular weight determination. *Journal of Biochemical and Biophysical Methods*, *60*(3), 281–293.

37. Gonzalez, I., Mestach, D., Leiza, J. R., & Asua, J. M., (2008). Adhesion enhancement in waterborne acrylic latex binders synthesized with phosphate methacrylate monomers. *Progress in Organic.*

38. Goodwin, J. W., Hearn, J., Ho, C. C., & Ottewill, R. H., (1973). The preparation and characterization of polymer latices formed in the absence of surface active agents. *British Polymer Journal*, *5*(5), 347–362.

39. Goto, A., & Fukuda, T., (2004). Kinetics of living radical polymerization. *Progress in Polymer Science*, *29*(4), 329–385.

40. Gugliotta, L. M., Vega, J. R., & Meira, G. R., (2000). Latex particle size distribution by dynamic light scattering: Computer evaluation of two alternative calculation paths. *Journal of Colloid and Interface Science*, *228*(1), 14–17.

41. Guillaume, J. L., Pichot, C., & Guillot, J., (1988). Emulsifier - free emulsion copolymerization of styrene and butyl acrylate. II. Kinetic studies in the presence of ionogenic comonomers. *Journal of Polymer Science Part A: Polymer Chemistry*, *26*(7), 1937–1959.

42. Hadjichristidis, N., Iatrou, H., Pispas, S., & Pitsikalis, M., (2000). Anionic polymerization: High vacuum techniques. *Journal of Polymer Science Part A: Polymer Chemistry*, *38*(18), 3211–3234.

43. Hayashi, S., Komatsu, A., & Hirai, T., (1989). Seeded polymerization of vinyl acetate using monodisperse poly (vinyl acetate) latex particles. *Journal of Polymer Science Part A: Polymer Chemistry*, *27*(1), 157–169.

44. Hearn, J., Wilkinson, M. C., Goodall, A. R., & Chainey, M., (1985). Kinetics of the surfactant - free emulsion polymerization of styrene: - The post nucleation stage. *Journal of Polymer Science: Polymer Chemistry Edition*, *23*(7), 1869–1883.

45. Held, A., & Mecking, S., (2000). Coordination polymerization in water affording amorphous polyethylenes. *Chemistry–A European Journal*, *6*(24), 4623–4629.

46. Herzog, C. A., Leite, C. A. P., & Galembeck, F., (1999). Latex particle self-assembly and particle microchemical symmetry: PS/HEMA latex particles are intrinsic dipoles. *Langmuir*, *15*(13), 4447–4453.
47. Ho, C. C., (2021). Fundamentals and recent applications of natural rubber latex in dipping technology. *Chemistry, Manufacture, and Application of Natural Rubber*, *4*, 317–361.
48. Horák, D., & Benedyk, N., (2004). Magnetic poly (glycidyl methacrylate) microspheres prepared by dispersion polymerization in the presence of electrostatically stabilized ferrofluids. *Journal of Polymer Science Part A: Polymer Chemistry*, *42*(22), 5827–5837.
49. Hua, L., Kai, W., Yang, J., & Inoue, Y., (2010). A new poly (L-lactide)-grafted graphite oxide composite: Facile synthesis, electrical properties, and crystallization behaviors. *Polymer Degradation and Stability*, *95*(12), 2619–2627.
50. Huang, Y., Wang, J., Zhou, J., Xu, L., Li, Z., Zhang, Y., & Jiang, L., (2011). Controllable synthesis of latex particles with multicavity structures. *Macromolecules*, *44*(8), 2404–2409.
51. Ikeda, S., & Nishinari, K., (2000). Intermolecular forces in bovine serum albumin solutions exhibiting solid like mechanical behaviors. *Biomacromolecules*, *1*(4), 757–763.
52. Imam, S. H., Mao, L., Chen, L., & Greene, R. V., (1999). Wood adhesive from crosslinked poly (vinyl alcohol) and partially gelatinized starch: Preparation and properties. *Starch - Stärke*, *51*(6), 225–229.
53. Jacob, J. L., D'Auzac, J., & Prevot, J. C., (1993). The composition of natural latex from Hevea Brasiliensis. *Clinical Reviews in Allergy*, *11*, 325–337.
54. Jayachandran, K. N., Takacs-Cox, A., & Brooks, D. E., (2002). Synthesis and characterization of polymer brushes of poly (N,N-dimethylacrylamide) from polystyrene latex by aqueous atom transfer radical polymerization. *Macromolecules*, *35*(11), 4247–4257.
55. Joensson, J. E., Hassander, H., & Toernell, B., (1994). Polymerization conditions and the development of a core-shell morphology in PMMA/PS latex particles. 1. Influence of initiator properties and mode of monomer addition. *Macromolecules*, *27*(7), 1932–1937.
56. Kablitz, C. D., & Urbanetz, N. A., (2007). Characterization of the film formation of the dry coating process. *European Journal of Pharmaceutics and Biopharmaceutics*, *67*(2), 449–457.
57. Keddie, J. L., Routh, A. F., Keddie, J. L., & Routh, A. F., (2010). An introduction to latex and the principles of colloidal stability. *Fundamentals of Latex Film Formation: Processes and Properties*, *5*, 1–26.

58. Khorassani, M., Afshar - Taromi, F., Mohseni, M., & Pourmahdian, S., (2009). The role of auxiliary monomers and emulsifiers on wet scrub resistance of various latex paints at different pigment volume concentrations. *Journal of Applied Polymer Science*, *113*(5), 3264–3268.
59. Kohut, A., Demchuk, Z., Kingsley, K., Voronov, S., & Voronov, A., (2018). Dual role of methyl-β-cyclodextrin in the emulsion polymerization of highly hydrophobic plant oil-based monomers with various unsaturations. *European Polymer Journal*, *108*, 322–328.
60. Kunkel, J. P., & Urban, M. W., (1993). Surface and interfacial FT–IR spectroscopic studies of latexes. VIII. The effect of particle and copolymer composition on surfactant exudation in styrene - n - butyl acrylate copolymer latex films. *Journal of Applied Polymer Science*, *50*(7), 1217–1223.
61. Lamb, D. J., Anstey, J. F., Fellows, C. M., Monteiro, M. J., & Gilbert, R. G., (2001). Modification of natural and artificial polymer colloids by "topology-controlled" emulsion polymerization. *Biomacromolecules*, *2*(2), 518–525.
62. Landfester, K., (2009). Miniemulsion polymerization and the structure of polymer and hybrid nanoparticles. *Angewandte Chemie International Edition*, *48*(25), 4488–4507.
63. Lee, D. I., (2005). The effects of latex coalescence and interfacial crosslinking on the mechanical properties of latex films. *Polymer*, *46*(4), 1287–1293.
64. Lichti, G., Gilbert, R. G., & Napper, D. H., (1983). The mechanisms of latex particle formation and growth in the emulsion polymerization of styrene using the surfactant sodium dodecyl sulfate. *Journal of Polymer Science: Polymer Chemistry Edition*, *21*(1), 269–291.
65. Liu, B., Wang, Y., Zhang, M., & Zhang, H., (2016). Initiator systems effect on particle coagulation and particle size distribution in one-step emulsion polymerization of styrene. *Polymers*, *8*(2), 55.
66. Manguian, M., Save, M., & Charleux, B., (2006). Batch emulsion polymerization of styrene stabilized by a hydrophilic macro - RAFT agent. *Macromolecular Rapid Communications*, *27*(6), 399–404.
67. Marcu, I., Daniels, E. S., Dimonie, V. L., Hagiopol, C., Roberts, J. E., & El-Aasser, M. S., (2003). Incorporation of alkoxysilanes into model latex systems: Vinyl copolymerization of vinyltriethoxysilane and n-butyl acrylate. *Macromolecules*, *36*(2), 328–332.
68. Martinsen, A., Skjåk-Bræk, G., Smidsrød, O., Zanetti, F., & Paoletti, S., (1991). Comparison of different methods for determination of molecular

weight and molecular weight distribution of alginates. *Carbohydrate Polymers*, *15*(2), 171–193.

69. Maxwell, I. A., Kurja, J., Van, D. G. H., German, A. L., & Morrison, B. R., (1992). Partial swelling of latex particles with monomers. *Die Makromolekulare Chemie*, *193*(8), 2049–2063.
70. Michaudel, Q., Kottisch, V., & Fors, B. P., (2017). Cationic polymerization: From photoinitiation to photocontrol. *Angewandte Chemie International Edition*, *56*(33), 9670–9679.
71. Mock, E. B., De Bruyn, H., Hawkett, B. S., Gilbert, R. G., & Zukoski, C. F., (2006). Synthesis of anisotropic nanoparticles by seeded emulsion polymerization. *Langmuir*, *22*(9), 4037–4043.
72. Mohamed, A., Ardyani, T., Bakar, S. A., Brown, P., Hollamby, M., Sagisaka, M., & Eastoe, J., (2016). Graphene-philic surfactants for nanocomposites in latex technology. *Advances in Colloid and Interface Science*, *230*, 54–69.
73. Mohapatra, S., & Nando, G. B., (2013). Chemical modification of natural rubber in the latex stage by grafting cardanol, a waste from the cashew industry and a renewable resource. *Industrial & Engineering Chemistry Research*, *52*(17), 5951–5957.
74. Morton, M., & Fetters, L. J., (1975). Anionic polymerization of vinyl monomers. *Rubber Chemistry and Technology*, *48*(3), 359–409.
75. Muroi, S., Hashimoto, H., & Hosoi, K., (1984). Morphology of core - shell latex particles. *Journal of Polymer Science: Polymer Chemistry Edition*, *22*(6), 1365–1372.
76. Nallamilli, T., Binks, B. P., Mani, E., & Basavaraj, M. G., (2015). Stabilization of Pickering emulsions with oppositely charged latex particles: Influence of various parameters and particle arrangement around droplets. *Langmuir*, *31*(41), 11200–11208.
77. Nenaah, E. G., & Ahmed, M. E., (2011). Antimicrobial activity of extracts and latex of *Calotropis procera* (Ait.) and synergistic effect with reference antimicrobials. *Research Journal of Medicinal Plant*, *5*(6), 706–716.
78. Otsu, T., (2000). Iniferter concept and living radical polymerization. *Journal of Polymer Science Part A: Polymer Chemistry*, *38*(12), 2121–2136.
79. Painter, P. C., Park, Y., Sobkowiak, M., & Coleman, M. M., (1990). Coal solubility and swelling. 2. Effect of hydrogen bonding on calculations of molecular weight from swelling measurements. *Energy & Fuels*, *4*(4), 384–393.

80. Pantazis, D., Chalari, I., & Hadjichristidis, N., (2003). Anionic polymerization of styrenic macromonomers. *Macromolecules, 36*(11), 3783–3785.

81. Papynov, E. K., Mayorov, V. Y., Palamarchuk, M. S., Bratskaya, S. Y., & Avramenko, V. A., (2013). Sol–gel synthesis of porous inorganic materials using "core–shell" latex particles as templates. *Journal of Sol-Gel Science and Technology, 68*(3), 374–386.

82. Pelley, A. J., & Tufenkji, N., (2008). Effect of particle size and natural organic matter on the migration of nano-and microscale latex particles in saturated porous media. *Journal of Colloid and Interface Science, 321*(1), 74–83.

83. Qiao, X. G., Dugas, P. Y., Charleux, B., Lansalot, M., & Bourgeat-Lami, E., (2015). Synthesis of multipod-like silica/polymer latex particles via nitroxide-mediated polymerization-induced self-assembly of amphiphilic block copolymers. *Macromolecules, 48*(3), 545–556.

84. Rawlings, J. B., & Ray, W. H., (1988). The modeling of batch and continuous emulsion polymerization reactors. Part I: Model formulation and sensitivity to parameters. *Polymer Engineering & Science, 28*(5), 237–256.

85. Santos, R. M., & Forcada, J., (1996). Synthesis and characterization of latex particles with acetal functionality. *Trends in Colloid and Interface Science 10*, 87–90.

86. Schoonbrood, H. A., & Asua, J. M., (1997). Reactive surfactants in heterophase polymerization. 9. Optimum surfmer behavior in emulsion polymerization. *Macromolecules, 30*(20), 6034–6041.

87. Smulders, W., & Monteiro, M. J., (2004). Seeded emulsion polymerization of block copolymer core– shell nanoparticles with controlled particle size and molecular weight distribution using xanthate-based RAFT polymerization. *Macromolecules, 37*(12), 4474–4483.

88. Stubbs, J. M., & Sundberg, D. C., (2008). The dynamics of morphology development in multiphase latex particles. *Progress in Organic Coatings, 61*(2–4), 156–165.

89. Suddaby, K. G., Haddleton, D. M., Hastings, J. J., Richards, S. N., & O'donnell, J. P., (1996). Catalytic chain transfer for molecular weight control in the emulsion polymerization of methyl methacrylate and methyl methacrylate– styrene. *Macromolecules, 29*(25), 8083–8091.

90. Szwarc, M., (2005). Living polymers and mechanisms of anionic polymerization. *Living Polymers and Mechanisms of Anionic Polymerization*, 1–177.

91. Thomas, R. R., Lloyd, K. G., Stika, K. M., Stephans, L. E., Magallanes, G. S., Dimonie, V. L., & El-Aasser, M. S., (2000). Low free energy surfaces using blends of fluorinated acrylic copolymer and hydrocarbon acrylic copolymer latexes. *Macromolecules*, *33*(23), 8828–8841.
92. Timpa, J. D., (1991). Application of universal calibration in gel permeation chromatography for molecular weight determinations of plant cell wall polymers: Cotton fiber. *Journal of Agricultural and Food Chemistry*, *39*(2), 270–275.
93. Touchard, V., Graillat, C., Boisson, C., D'Agosto, F., & Spitz, R., (2004). Use of a Lewis acid surfactant combined catalyst in cationic polymerization in miniemulsion: Apparent and hidden initiators. *Macromolecules*, *37*(9), 3136–3142.
94. Ukrainczyk, N., & Rogina, A., (2013). Styrene–butadiene latex modified calcium aluminate cement mortar. *Cement and Concrete Composites*, *41*, 16–23.
95. Vega, J. R., Gugliotta, L. M., Gonzalez, V. D., & Meira, G. R., (2003). Latex particle size distribution by dynamic light scattering: Novel data processing for multiangle measurements. *Journal of Colloid and Interface Science*, *261*(1), 74–81.
96. Vignati, E., Piazza, R., & Lockhart, T. P., (2003). Pickering emulsions: Interfacial tension, colloidal layer morphology, and trapped-particle motion. *Langmuir*, *19*(17), 6650–6656.
97. Voorn, D. J., Ming, W., & Van, H. A. M., (2006). Polymer− clay nanocomposite latex particles by inverse pickering emulsion polymerization stabilized with hydrophobic montmorillonite platelets. *Macromolecules*, *39*(6), 2137–2143.
98. Wang, J. S., & Matyjaszewski, K., (1995). Controlled/" living" radical polymerization. atom transfer radical polymerization in the presence of transition-metal complexes. *Journal of the American Chemical Society*, *117*(20), 5614, 5615.
99. Wedlock, D. J., Fasihuddin, B. A., & Phillips, G. O., (1986). Comparison of molecular weight determination of sodium alginate by sedimentation-diffusion and light scattering. *International Journal of Biological Macromolecules*, *8*(1), 57–61.
100. Wen, G., Zhang, Y., Zhang, Y., Sun, K., & Fan, Y., (2002). Rheological characterization of storage-stable SBS-modified asphalts. *Polymer Testing*, *21*(3), 295–302.
101. Wu, C. H., Don, T. M., & Chiu, W. Y., (2011). Characterization and conversion determination of stable PEDOT latex nanoparticles synthesized by emulsion polymerization. *Polymer*, *52*(6), 1375–1384.

102. Yağci, Y., & Reetz, I., (1998). Externally stimulated initiator systems for cationic polymerization. *Progress in Polymer Science*, *23*(8), 1485–1538.
103. Yan, X., Ji, Y., & He, T., (2013). Synthesis of fiber crosslinking cationic latex and its effect on surface properties of paper. *Progress in Organic Coatings*, *76*(1), 11–16.
104. Zhang, Y., Wang, J., Huang, Y., Song, Y., & Jiang, L., (2011). Fabrication of functional colloidal photonic crystals based on well-designed latex particles. *Journal of Materials Chemistry*, *21*(37), 14113–14126.

CHAPTER

3 Physics and Mechanics of Latex

Contents

3.1. INTRODUCTION

The domain of latex, along with latex bonding, includes a broad spectrum of applications and sectors, from production to healthcare and various other related fields. Latex is a colloid type consisting of polymer particles dispersed in a liquid medium. Its unique properties render it a versatile and valuable material in diverse applications. To gain a comprehensive understanding and effectively utilize latex-based products in order it is imperative to delve through the fundamental principles of physics along with mechanisms that dictate their behavior. Comprehending the underlying principles that regulate the characteristics and actions of latex particles, including their molecular configuration, composition, and rheological behaviors, is integral to studying latex physics. Latex particles are known for their distinctive characteristics that stem from their molecular configuration and intermolecular forces. These particles are typically made of synthetic polymers such as styrene-butadiene or organic, including natural rubber (Lepizzera et al., 1997; Wang et al., 2006).

The rheological properties of latex are a fundamental characteristic that pertains to its response to external stress and deformation. Its viscosity determines the flow properties of latex, whereas its ability to deform and regain its original shape is attributed to its elasticity. The characteristics above are of paramount importance in various industrial procedures such as coating, adhesion, and molding, in which the fluidity and malleability of latex are integral factors. The process of latex film creation is a multifaceted phenomenon that encompasses converting a scattered latex system through a cohesive and uninterrupted film. Comprehending the underlying mechanisms of this process is crucial to enhance the formation of films and attaining the intended characteristics in products that are based on latex. Various factors, including pH, temperature, and additives, influence the destabilization and reaction of coagulation for the latex particles. These factors contribute to the aggregation of particles and subsequent film growth (Hu et al., 1996; Xie et al., 1997).

The adhesion of latex films is contingent upon distinct mechanisms and interactions occurring at the interface linking the latex and the substrate. Latex bonding has been used in diverse industrial sectors, such as biomedical applications, adhesive formulations, and textile coatings. Investigating the physics underlying latex bonding provides insight into the interfacial phenomenon and adhesion mechanisms at play, facilitating the formulation of efficacious bonding approaches. Physical interactions significantly influence the bonding of latex. Upon contact between a latex film along with the substrate, intermolecular interactions, including van der Waals forces along with hydrogen bonding, are activated. The forces above facilitate the adhesion of the latex film onto the substrate, thereby conferring mechanical robustness and longevity. Comprehending the physical mechanisms underlying latex bonding facilitates the development of enhanced bonding methodologies and identifying suitable substrates (Wicks et al., 2006; Poompradub et al., 2014).

The evaluation of the mechanical characteristics of latex films is crucial in ascertaining their efficacy in diverse usage scenarios. The mechanical properties of latex-based materials, such as tensile strength and elongation, are crucial in determining their durability, flexibility, and ability to resist deformation. The utilization of suitable testing techniques to assess these characteristics offers valuable understanding regarding the response of latex films to varying stress scenarios, thereby facilitating the enhancement of their efficacy. In recent times, the progressions in latex technological advances and new developments have presented novel opportunities for material design along with applications. Notwithstanding, there are still several challenges and constraints that need to be addressed, including the optimization of film formation, the improvement of bonding strength, and the resolution of environmental issues related to latex-based commodities. Through a concentrated effort on these obstacles, scholars and practitioners can pinpoint domains that warrant additional investigation and advancement, resulting in forthcoming advancements and discoveries (Quach, 1973).

In summary, a comprehensive comprehension of the physics and processes that underlie latex along with latex bonding is crucial for fully exploiting their potential. By understanding the characteristics, mechanisms of film development, bonding processes, and mechanical characteristics of materials based on latex, researchers, and practitioners can devise enhanced materials and more efficient bonding methodologies and investigate innovative applications. The forthcoming chapter will comprehensively examine the physics and processes that govern latex and latex relationships, thereby facilitating advancements in this intriguing domain (Zhang et al., 2002).

3.2. LATEX PHYSICS

Latex is a type of colloid comprising polymer particles dispersed within a liquid medium. Its distinct characteristics and adaptability make it a frequently employed material across diverse fields. Latex can be obtained from either natural or synthetic polymers, thereby providing a diverse array of material alternatives to cater to various industrial requirements. Latex possesses several noteworthy characteristics that render it immensely valuable across diverse disciplines. Primarily, latex particles exhibit a spherical morphology, and their dimensions range from nanometers to micrometers, contingent upon the precise composition. The distribution of particle sizes plays a significant role in determining latex's colloidal integrity and flow characteristics. Furthermore, latex exhibits notable resistance to coagulation or settling, rendering it capable of enduring prolonged storage durations without substantial alterations in its characteristics (Santipanusopon & Riyajan, 2009).

Latex possesses a crucial characteristic of being capable of undergoing film formation. Upon application of a latex dispersion over a substrate or exposure to specific circumstances, the solvent undergoes evaporation, thereby causing the coalescence and merging of latex particles, culminating in the creation of a coherent film. The film-forming characteristic of latex renders it suitable for deployment in various applications, including coatings, adhesives, and diverse other industrial and commercial uses (Teixeira & Bon, 2010; Shaffie et al., 2018).

3.2.1. Molecular Structure and Composition of Latex

The molecular configuration of latex particles depends on the specific polymer utilized. The principal polymer found in natural rubber latex is poly(cis-1,4-isoprene). In contrast, synthetic latexes comprise polymers that include styrene-butadiene, acrylics, and polyvinyl acetate. The mechanical and physical characteristics of the resultant latex films are influenced by the molecular makeup of these polymers. In a latex particulate, the chains of polymers are commonly intertwined and convoluted as a result of Brownian motion along with thermal agitation. The entanglement phenomenon creates a structure that exists in three dimensions, forming a network. The mechanical properties of latex films, such as tensile strength, elongation, and flexibility, are affected by the degree of entanglement alongside the length of the chain of polymers (Almeida et al., 2000).

Latex is comprised of not only polymers but also a variety of stabilizers and additives. The additives mentioned above are essential in regulating the stability of the rubber dispersion, inhibiting particle aggregation, and guaranteeing homogeneous film formation. Incorporating additives into a given system may encompass a range of compounds, such as surfactants, dispersants, pH modifiers, and cross-linking agents, contingent upon the particular utilization and intended characteristics.

3.2.2. Rheological Behavior of Latex

The rheological characteristics of latex pertain to its response to external stress and deformation. The behavior above holds significant importance in various processes, including coating, shaping, and adhesion, wherein the flow and deformation of latex are crucial determinants. Its viscosity determines the flow properties of latex. The phenomenon refers to the resistance of a liquid medium to flow, which is subject to various factors, including but not limited to the concentration of polymers, particle size distribution, and the existence of added substances. Latex dispersions display non-Newtonian characteristics, whereby their viscosity depends on the shear rate used. Frequently observed phenomena in fluid mechanics involve shear-thinning behavior, which is characterized by a reduction in viscosity to be the rate at which shear occurs increases, and shear-thickening conduct, which is characterized by a rise in viscosity as the power source shear rate rises (Shaffie et al., 2018) (Figure 3.1).

Tip

The rheological behavior of latex, such as its viscosity and flow properties, plays a significant role in its processability and the quality of coatings or films produced from it.

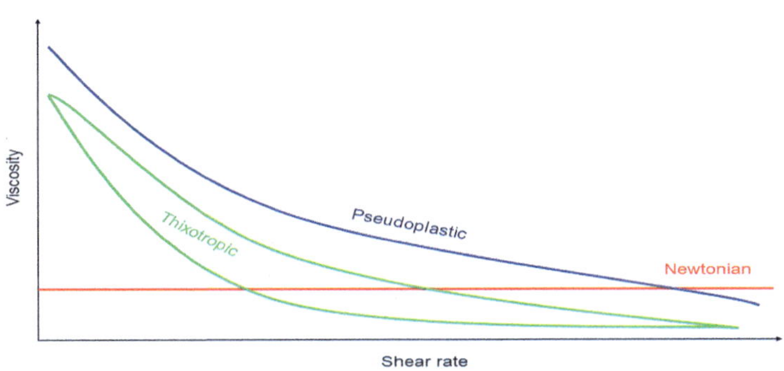

Figure 3.1. Rheological flow behavior of different fluids.

Source: Artur Palasz, Creative Commons License.

The rheological characteristic known as elasticity holds significant importance in latex. The term pertains to the capacity of the latex paint film to undergo deformation in response to stress and subsequently recover its initial shape upon removal of the stress. The elastic properties of latex particles result from the intermingling and structural organization of polymer chains inside them. The degree of elasticity present in latex films has a significant impact on their mobility, resilience, and capacity to endure deformation. The distinctive behavior of latex is attributed to its viscoelastic characteristics, which encompass features of both elasticity and viscosity. The behavior mentioned above facilitates the ability of latex to conform to diverse shapes throughout coating or molding procedures while simultaneously conferring the requisite mechanical robustness and steadfastness in the resultant films.

3.3. LATEX FILM FORMATION MECHANISMS

The procedure of latex film creation is intricate and involves converting a scattered latex structure into a cohesive and uninterrupted film. Comprehending the mechanisms that underlie this process is of paramount importance in the optimization of film formation and the attainment of desired properties in products based on latex. The mechanisms of film formation are influenced by many variables, which include along with outside variables like temperature, pH, along with additives (Steward et al., 2000; Dobler & Holl, 1996) (Figure 3.2).

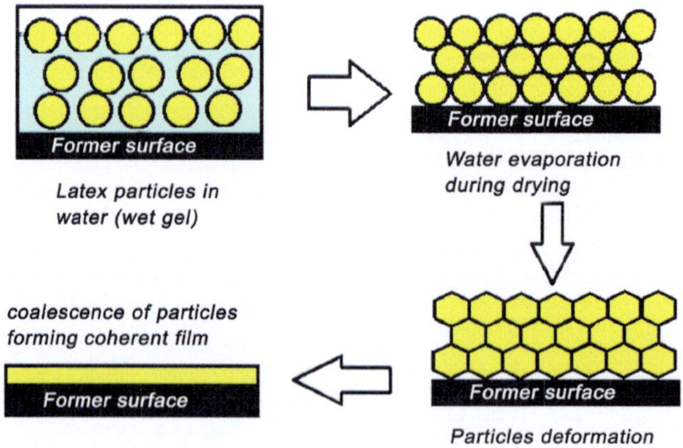

Figure 3.2. Schematic diagram of latex film formation.

Source: Nurul Huda, Creative Commons License.

3.3.1. Latex Coagulation and Destabilization Processes

The coagulation phenomenon pertains to the amalgamation of latex particles resulting in the formation of comparatively larger aggregates. Latex coagulation may arise from diverse mechanisms such as chemical reactions, alterations in temperature, or the introduction of coagulating agents. The process of coagulation may be advantageous in specific contexts, such as in the case of adhesives that are based on latex, as it facilitates rapid bonding. It is imperative to prevent coagulation in various applications, including coatings, to uphold a consistent dispersion (Winnik, 1997).

In contrast, destabilization mechanisms refer to disrupting the colloidal equilibrium of latex particles, resulting in their coalescence and subsequent film formation. The destabilization phenomenon can be brought about by various mechanisms, including but not limited to Ostwald ripening, flocculation, and depletion flocculation. The phenomenon of Ostwald ripening entails the progressive dissolution of diminutive latex particles followed by the repositioning of the disintegrated polymer onto larger particles, culminating in the expansion of larger aggregates. The phenomenon of flocculation takes place when the destabilizing forces surpass the repulsive forces among the particles of latex, leading to their aggregation and the formation of flocs. The process of depletion flocculation entails the introduction of a non-adsorbing agent, such as a polymer or surfactant, which triggers attractive forces among latex particles. This phenomenon ultimately results in the formation of flocculates (Routh & Russel, 2001; Vanderhoff, 1970).

3.3.2. Particle Aggregation and Film Growth Mechanisms

After undergoing coagulation or instability, latex particles agglomerate and merge, forming a cohesive film. The process of aggregation, along with film growth, encompasses multiple mechanisms. Initially, it should be noted that when latex particles are in close contact, interparticle forces, which include van der Waals forces, become operative. The cohesive forces play a significant role in the agglomeration of particles, facilitating their adherence and amalgamation (Henson et al., 1953).

The process of particle aggregation is subject to additional factors, including the existence of stabilizing agents and supplementary substances within the latex system. Stabilizing agents, which include surfactants or dispersants, serve to mitigate or regulate particle agglomeration, thereby promoting homogeneous dispersion and facilitating film formation. Furthermore, the inclusion may enhance the inter-particle bonding of latex particles, resulting in the development a more unified film (Dragnevski & Donald, 2008) (Figure 3.3).

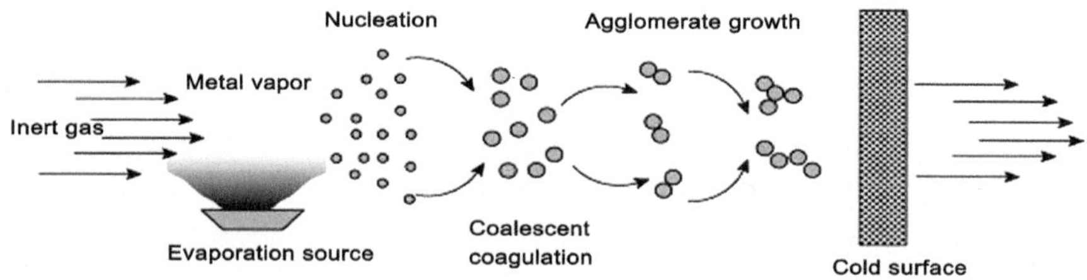

Figure 3.3. Schematic diagram of the processes that contribute to particle growth.

Source: Ismat Shah, Creative Commons License.

Film growth entails the reorganization and dispersion of latex particles, leading to the development of a cohesive and uninterrupted film architecture. The process of latex particle fusion occurs as a result of the gradual evaporation of the liquid medium within the latex dispersion, which causes the particles to move closer together and ultimately merge with adjacent particles.

The amalgamation of particles is facilitated by various mechanisms, including but not limited to particle coalescence, which involves the merging of adjacent particles, and the dispersion of chains of polymers across particle user interfaces. The film growth process involves the liquid medium's evaporated state, which subsequently causes an elevation in the latex solids materials.

This phenomenon ultimately leads to the development of a film structure that is more densely packed. The ultimate thickness and consistency of the film are impacted by several factors, including the initial viscosity of the latex dispersion, the speed at which the solvent evaporates, and the extent of aggregation among the latex particles (Sheetz, 1965; Routh & Russel, 1999).

3.3.3. Factors Influencing Latex Film Formation

Several external factors significantly influence the process of latex film formation. The film formation process and the resultant film's properties can be significantly impacted by factors such as the pH level, temperature, and the number of additives. The process of latex film formation is subject to the influence of temperature, which modulates the rate of solvent loss. Elevated temperatures expedite the process of solvent evaporation, thereby resulting in an accelerated rate of film formation. Nevertheless, elevated temperature can lead to unfavorable consequences, including coagulation or creation of non-homogeneous films. The stability of latex and its particle interactions are influenced by ph. Variations in pH levels can potentially modify the surface electrical charge of latex, petroleum, and particles, thereby influencing their repellent characteristics and consequently affecting the film formation process (Keddie, 1997).

Latex formulations are frequently augmented with additives, including surfactants, plasticizers, or cross-linking agents, to alter film characteristics and improve the process of film formation. Incorporating plasticizers into a film matrix has been found to enhance its flexibility, whereas the addition of surfactants has been shown to improve the stability of latex dispersion. Cross-linking agents facilitate the adhesion between latex particles, leading to a more unified film structure that exhibits improved mechanical robustness. In addition, the selection of substrates or surface properties may impact the formation of latex films. The wetting, along with the adhesion of the latex film, can be influenced by & the substrate, resulting in fluctuations in film thickness and morphology.

3.4. LATEX BONDING MECHANISMS

The process of latex bonding entails the attachment of rubber films to surfaces, resulting in a robust and long-lasting adhesion. This substance is extensively utilized across diverse sectors, such as textiles, adhesives, biomedical domains, and coatings. Comprehending the mechanisms underlying latex bonding is imperative in developing productive bonding tactics and enhancing the functionality of materials based on latex. The process of latex bonding presents several benefits, such as its adaptability, flexibility, and ability to harmonize with various underlying materials. This technique facilitates the adhesion of dissimilar materials, including plastics, metals, clothes, and biological substrates. The adaptability of latex bonding renders it a valuable methodology in industries that necessitate the amalgamation of diverse materials (Selomulya et al., 2002; Tian et al., 2021) (Figure 3.4).

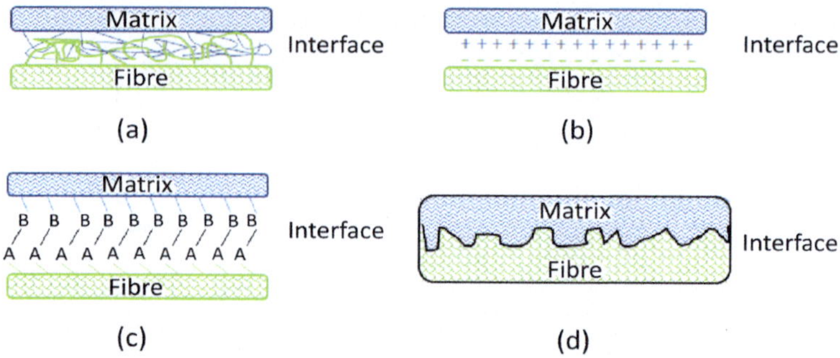

Figure 3.4. Latex-matrix interfacial bonding mechanism.

Source: Yonghui Zhou, Creative Commons License.

3.4.1. Interfacial Phenomena and Adhesion in Latex Bonding

The efficacy of latex bonding is contingent upon establishing robust interface adhesion between the backing material and the latex film. The role of interfacial issues in bonding is paramount, as it governs the adhesive joint's durability and quality. Upon contact between a latex film, relationships take place at the interface. The various types of interactions between molecules encompass hydrogen bonding, van der Waals forces, & electrostatic forces, in specific instances, covalent bonding. The characteristics and intensity of these interactions are contingent upon the particular materials engaged and the surface attributes (Stevens et al., 2007). The bonding of latex is typically classified into two primary categories, namely mechanical attachment and chemical adhesion. The phenomenon of mechanical adhesion pertains to the physical interconnected or getting caught of latex strands with the outermost layer of the substrate, leading to the formation of a robust mechanical bond. Chemical attachment is a phenomenon that arises from the chemical reactions that occur among the functional groups present upon the particles of latex and the surface of the substrate. This interaction leads to the formation of covalent or polar relationships, which in turn, augment the strength of the bond (Shahani et al., 1999).

3.4.2. Physical Interactions Contributing to Latex Bonding

The adhesion that occurs between the latex film along with the substrate is significantly influenced by physical interactions that are involved in latex bonding. The interactions, as mentioned earlier, encompass:

1. Van der Waals forces refer to feeble intermolecular forces that emerge from momentary fluctuations in the distribution of electrons. These entities' presence may enhance the latex film's adhesive properties by promoting attraction with the substrate (Wang et al., 2020).
2. Electrostatic forces are generated due to the interaction between charged particles, resulting in either attraction or repulsion. When either the particles of latex or the surface of the substrate possess a net responsibility, electrostatic attraction may be involved, thereby augmenting adhesion (Pan et al., 2020).
3. Hydrogen bonding is when hydrogen atoms within a given molecule engage with electronegative atoms present in a separate molecule. The phenomenon of hydrogen bonding has the potential to augment the adhesion within latex bonding, particularly in cases where either the latex particles along with the substrate possess functional groups that are capable of participating in hydrogen bonding (Pan et al., 2016).
4. Mechanical interconnected is a phenomenon observed in latex bonding, wherein the physical integration of latex strands or parts occurs through the deviations or roughness visible on the substrate's surface. The interlocking mechanism serves to augment the adhesion properties of the system by enlarging the surface area of contact that exists between the rubber film and the substrate's surface.

Did You Know?

The physical interactions involved in latex bonding include van der Waals forces, electrostatic forces, and entanglement of polymer chains, which contribute to the adhesion and strength of bonded latex surfaces.

The efficacy and potency of these chemical reactions are contingent upon variables such as surface texture, surface tension, and the congruity between the latex and underlying materials.

The interfacial relationships and bonding achievement can be enhanced by employing surface-preparing techniques, which involve substrate treatment, roughening, alongside cleaning (Poovaneshvaran et al., 2020).

3.5. MECHANICAL PROPERTIES OF LATEX FILMS

The assessment of mechanical properties is of paramount importance in determining the efficacy and appropriateness of latex films for diverse applications. Comprehending the mechanical properties of latex films is imperative to enhance their design, guarantee longevity, and fulfill particular performance criteria. The present section centers on two fundamental mechanical characteristics: tensile strength and elongation. It delves into the factors that affect their behavior and expounds on the testing techniques employed to assess these properties (Zosel & Ley, 1993).

3.5.1. Tensile Strength and Elongation of Latex Films

Tensile strength refers to the capacity of a substance to resist tensile stress, which is the force applied per unit area of a cross-section perpendicular to the applied force without experiencing rupture or failure. The maximum stress a material can endure before fracturing is a crucial parameter that measures a film's durability and tensile strength. Latex films exhibiting elevated tensile strength can endure external forces and exhibit resilience against deformation or failure. "elongation" pertains to the degree of deformation or stretching a substance can undergo before its rupture. The parameter referred to is the film's capacity to experience plastic deformation without undergoing fracture. The elongation of a film is commonly quantified as a percentage, which denotes the extent of elongation relative to the initial length of the film. Latex films exhibiting elevated elongation values possess enhanced flexibility and can endure greater deformations before experiencing failure (Zosel, 1995; Lee, 2005). The strength and length of latex polymer are influenced by several factors, such as the molecular makeup of the polymer, the number of fillers or ingredients, and the method of creating the film. The mechanical characteristics of films are influenced by the selection of latex polymers, which may include natural rubber and synthetic substances such as styrene-butadiene and acrylics. The strength and flexibility of polymers vary, resulting in differences in yield strength and elongation. The mechanical characteristics of the resultant films are also influenced by the nature and concentration of additives or fillers integrated into the latex formulation. Incorporating plasticizers into a polymer matrix can enhance the flexibility and elongation of films by mitigating intermolecular forces. The incorporation of reinforcing agents such as fillers as well as fibers can augment the tensile strength of a material by imparting supplementary structural reinforcement (Figure 3.5).

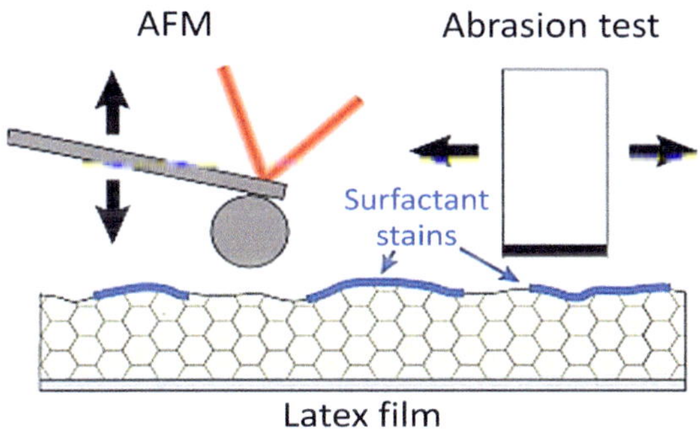

Figure 3.5. Robustness of surfactant-laden latex films.

Source: Wojciech Kwieciński, Creative Commons License.

3.5.2. Factors Influencing the Mechanical Behavior of Latex Films

The mechanical properties of latex films are influenced by multiple factors:

1. The mechanical characteristics of latex are directly influenced by the selection of latex polymer along with its molecular weight. Films from polymers with greater molecular weights typically exhibit elevated tensile strength, albeit reduced elongation.
2. The mechanical characteristics of a latex film are affected by the number of agents that cross-link and the extent of cross-linking. The process of cross-linking improves the tensile strength of a material by creating a networked structure that can withstand deformation. However, it is possible that cross-linking could hurt elongation (Winnik, 1997).
3. Its thickness influences the mechanical properties of latex film. Films with greater thickness typically demonstrate elevated tensile strength due to amplified material volume. In contrast, films with lesser thickness may exhibit greater lengthening due to reduced resistance to deformations.
4. The process of curing or drying is a crucial factor in the enhancement of mechanical properties. Adequate cross-linking and chain polymer getting caught can be achieved through appropriate curing or drying conditions, resulting in enhanced tensile strength and elongation.

5. The mechanical characteristics of latex films may be influenced by various environmental factors, such as temperature and humidity, along with contact with chemicals or UV radiation, which can cause changes over a period of time. Degradation may occur due to environmental factors, resulting in a decrease in both tensile strength and elongation (Manroshan & Baharin, 2005; Xu et al., 2022).

3.5.3. Testing Methods for Evaluating Mechanical Properties

Multiple testing techniques are employed to assess the mechanical characteristics of latex films:

1. The measurement of the elongation and tensile strength is commonly conducted using a tensile testing apparatus. The latex film samples undergo a regulated tensile force until rupture, and the resultant stress-strain graph furnishes data about tensile potency, elongation, and additional mechanical parameters (Charmeau et al., 1997).
2. Microscopy is a valuable tool for analyzing the surface structure and microstructural characteristics of latex films. SEM, or scanning electron microscopy, is one microscopic technique that can be employed for this purpose. This aids in evaluating variables such as the dispersion of particles, uniformity of the film, and imperfections or irregularities that could potentially impact the mechanical characteristics (Zhong & Chen, 2002).
3. Dynamic Mechanically Analysis (DMA) is a technique utilized to investigate the viscoelastic properties of latex films when subjected to dynamic loading conditions. The data above of storage modulus, lack modulus, and damping properties are significant in comprehending the film's reaction to varying frequencies and temperatures (McMahan et al., 2005; Joanicot et al., 1996).
4. The evaluation of a film's capacity to withstand abrupt and high-energy impacts is conducted through impact testing. Several impact testing techniques, including the Izod or the Charpy tests, can be utilized to assess the film's durability and ability to withstand fracture (Lepizzera et al., 1997; Misra et al., 1983).
5. Peel or adhesion methods of testing may be employed to evaluate the robustness and endurance of the bond that exists between the

latex film and the backing material is bonded latex assemblies. The tests entail the application of a regulated force to disengage the adhered surfaces and quantify the force needed (Roslim et al., 2018; Limousin et al., 2019; Afiq & Azura, 2013).

To summarize, assessing elongation and tensile strength mechanical properties is crucial in determining the effectiveness and appropriateness of latex films for various uses. Several factors, such as the composition of latex, cross-linking, the thickness of the film, curing conditions, and environmental factors, impact these properties. Various testing techniques, including impact testing, tensile testing, microscopy, and DMA, along with peel or attachment testing, offer valuable insights into the mechanical properties of latex films. These methods can be utilized to enhance the design of these films for targeted applications (Wang et al., 2016; Norhazariah et al., 2016).

3.6. FACTORS INFLUENCING LATEX BONDING STRENGTH

The strength of latex bonding is a crucial aspect in guaranteeing the longevity and dependability of adhesive bonds and coatings. The bonding strength among latex films along with substrates in order is influenced by various factors. This section delves into the various factors associated with surface preparation methods, chemical reactions, and the significance of curing and drying in latex bonding (Fathy et al., 2022).

3.6.1. Surface Preparation Techniques for Latex Bonding

Adequate surface preparation is imperative to attain robust and long-lasting latex adhesion.

Remember

When aiming to achieve strong latex bonding, it is essential to implement effective surface preparation techniques, such as cleaning, roughening, or applying primers, to enhance the adhesion between latex surfaces.

The objective is to generate a pristine and adequately primed substrate interface that facilitates adherence and augments cohesive potency. Various surface preparation techniques can be employed for latex bonding:

1. The process of cleaning the surface of the substrate is of utmost importance as it aids in eliminating any impurities, including but

not limited to debris, dust, oils, or grease. Frequently employed techniques for cleaning surfaces encompass the utilization of solvent scrubbing, degreasing, or surface-active agents. The presence of clean surfaces facilitates enhanced wetting and interaction between the latex film along with the substrate, thereby resulting in heightened bonding strength (Ortega-Vinuesa et al., 2001; Bluestein et al., 1975).

2. The surface roughening process improves the mechanical interlocking between the substrate and the latex film by enhancing their adhesion. Various methods, including sanding, abrasion, and etching, can generate a surface with enhanced roughness, thereby increasing the area of the surface and attachment sites (Magbool & Tayeh, 2021).
3. Surface activation methods are utilized to alter the surface of a substrate to improve bonding interactions. Techniques such as plasma treatment, corona treatment, and flame treatment have the potential to enhance surface energy, augment wettability, and stimulate chemical bonding between the latex film along with the substrate.
4. The application of a primer layer on the surface of the substrate before latex bonding has the potential to improve adhesion by creating a cohesive adhesive layer that is compatible with both the substrate and the latex film. The application of primers has been shown to enhance wetting properties, augment chemical compatibility, and facilitate interfacial bonding (Xu et al., 2006; Rith et al., 2016).
5. The present study investigates the chemical interactions contributing to latex bonding.
6. The bonding of latex is significantly influenced by chemical interactions occurring at the interface between the substrate or the latex film. The interactions above have the potential to augment adhesion and thereby bolster the bonding efficacy. Crucial chemical interactions encompass.
7. Van der Waals forces, which result from brief changes in electron distribution, play a role in the adhesion between the latex mattresses film and the backing material. The forces mentioned above exert a noteworthy influence in facilitating adhesion and augmenting the strength of bonding.

8. Hydrogen bonding is a phenomenon that takes place when hydrogen atoms within a particular molecule engage with electronegative atoms present in a different molecule. Hydrogen bonding can augment adhesion within latex bonding, particularly in cases where the particles of latex or the substrate possess functional groups capable of participating in hydrogen bonding.
9. Covalent bonding can occur among functional structures on latex particles and the substrate, resulting in robust and long-lasting adhesive joints in certain instances. The present bonding mechanism exhibits exceptional adhesion properties and demonstrates resilience in challenging environmental conditions (Shunmuga et al., 2023; Xu et al., 2016).

The chemical interactions that are responsible for latex bonding are contingent upon several factors, including the type of latex polymer, the functional groups that are present within latex parts, and the substrate materials.

3.6.2. Role of Curing and Drying in Latex Bonding

The processes of curing and drying plays a crucial role in the development of the desired characteristics and bonding strength within latex bonding. The term "curing" pertains to the biological or chemical mechanisms that facilitate the cross-linking and the polymer chain caught throughout the latex film, thereby augmenting the film's durability and strength. The drying process entails eliminating the solution from the rubber film, which facilitates the development of a cohesive and uninterrupted film. Throughout the curing process, chemical processes or physical phenomena, such as radiation, heat, or moisture, facilitate the development of chemical bonds and intermolecular interactions in the latex film. The cross-linking reactions can be facilitated by the addition of cross-linking agents as well catalysts. Adequate cross-linking and attainment of optimal bonding capacity necessitate adherence to appropriate curing conditions and adequate duration (Lee et al., 2016).

The desiccation process of the latex film holds equal significance, as it enables the latex particles to establish closer proximity and promotes amalgamation and interparticle adhesion by eliminating the liquid medium. Optimal drying conditions, encompassing factors such as humidity, temperature, along with airflow, facilitate consistent and regulated drying, thereby resulting in enhanced film formation, and strengthened bonding efficacy.

3.7. FUTURE DIRECTIONS AND EMERGING TRENDS

Despite the considerable progress and extensive application of latex and latex bonding, researchers and engineers encounter various challenges and constraints. Several contemporary obstacles are currently being faced:

1. The optimization of bonding strength is a persistent challenge, despite advancements in comprehending latex bonding processes. The attainment of uniform high bonding power across diverse substrates and environmental circumstances remains elusive. Pursuing to augment adhesive efficacy while preserving flexibility and resilience is a continuous subject of scholarly investigation (Gill et al., 2022).
2. The significance of environmental considerations in latex and latex connecting research cannot be overstated, as with any other material. Future development in the field of latex products is contingent upon the advancement of environmentally conscious formulations, the mitigation of the presence of VOCs within latex goods, and the exploration of renewable raw materials (Vansteenkiste et al., 2010).
3. The compatibility of latex bonding alongside difficult substrates may be hindered due to the low surface-energy content or chemically inert surfaces. Enhancing the bonding efficacy on these complex substrates is currently a thriving field of investigation (Goyal et al., 2019).
4. Further investigation is required to comprehend the aging behavior and durability of latex films and bonded assemblies. The performance and durability of latex-based content can be influenced by various factors, including chemical environments, temperature, UV exposure, and humidity. Comprehending and mitigating the impacts of aging is crucial for broadening the scope of potential applications for latex bonding (Shankar & Hollinger, 2007).

3.7.1. Advances in Latex Technology and Potential Applications

The progressions in latex technology have facilitated novel opportunities and broadened the scope of feasible applications. Several significant breakthroughs and prospective uses have been identified:

1. The incorporation of responsive polymers, nanoparticles, or functional additives into latex formulas has resulted in

the emergence of intelligent latex materials. The materials above possess characteristics such as autonomous mending, responsiveness to stimuli, and regulated dispensation, thereby creating opportunities for utilization in sensor technology, pharmaceutical transport mechanisms, and sophisticated coatings (Vansteenkiste & Mouratidis, 2016).

2. The scientific community is currently investigating the creation of latex formulations that are both biocompatible as well as biodegradable, intending to utilize them in a range of biomedical contexts. The versatility and biocompatibility of these materials make them suitable for various healthcare applications such as drug delivery systems, surgical adhesives, wound dressings, and tissue engineering (Kuppusamy et al., 2017).
3. Recent advancements in latex technology have facilitated the creation of high-performance coatings that exhibit enhanced chemical resistance, physical characteristics, and durability. The coatings mentioned above are utilized in various industries, including but not limited to transportation, aviation, and infrastructure, as they provide benefits such as corrosion prevention, weather endurance, and visual enhancement (Ludlow et al., 1999).
4. Researchers are currently exploring the integration of electronic parts into latex formulas as a means of developing electronic devices that are both flexible and stretchable. Utilizing Latex-based electronics exhibits promising prospects in wearable electronics, flexible displays, and electronic skin, facilitating novel functionalities and design prospects (Ogbebor et al., 2020).

3.7.2. Potential Areas for Further Research and Development

There exist various domains that present prospects for additional investigation and advancement in the realm of latex and latex bonding:

1. Exploring innovative bonding methods and surface modification strategies, including novel adhesive chemistries and Nanostructuring, can enhance bonding strength and broaden the scope of bonding substrates.
2. The exploration of integrating supplementary functionalities, which include antimicrobial, self-cleaning, or optical characteristics, into latex compositions can result in the creation

of multifunctional materials that exhibit improved performance and adaptability (Dralle et al., 2012).

3. The development of sustainable and environmentally friendly latex formulations through the utilization of renewable resources, bio-based monomers, and the reduction of the negative environmental effects of latex manufacturing and waste management is imperative for the industry's longevity (Cegala, 2006).
4. The advancement of characterization techniques, including in situ analysis throughout spectroscopic methods, film formation, and high-resolution microscopy, and a more profound understanding of the mechanisms and structure-property correlations in latex films and bonded meetings.

By exploring these research domains and resolving existing obstacles, the utilization of latex along with latex bonding may be expanded, thereby fostering progress in many sectors, such as healthcare, gadgets, coatings, and beyond.

Activity 3.1

In a research and development laboratory focused on latex materials, the team aims to enhance the bonding strength of a latex film. They would conduct experiments to evaluate and improve the bonding strength by optimizing surface preparation techniques, curing, and drying processes, and utilizing appropriate testing methods for mechanical properties. The effectiveness of these optimizations would be assessed through measurements and tests to ensure superior bonding strength in the latex film.

3.8. SUMMARY

The chapter titled "Physics along with Mechanisms of Latex along with Latex Bonding" explores the basic concepts and various factors that affect the behavior of latex, as well as the bonding strength among latex films along with substrates. The chapter underscores the significance of comprehending the physics along with mechanisms underlying latex along with latex bonding to enhance material properties and formulate efficacious bonding approaches. The introductory section of the chapter highlights the importance of investigating the physics of latex and the mechanisms of bonding. The chapter's contents are delineated, furnishing a guide for the ensuing sections. The segment about latex physics delves into the characterization and attributes of latex, encompassing its colloidal properties and capacity to form films. The text delves into the molecular architecture and constituent elements of latex, along with its rheological characteristics, encompassing viscosity, and elasticity (Mazarji et al., 2021; Nalla & Gurinckaya, 2020; Koyuncu et al., 2017).

Subsequently, the chapter explores the mechanisms underlying the formation of latex films. The subject matter pertains to the processes of latex coagulation and destabilization, as well as the mechanisms of particle aggregation and film growth. This section elucidates the various factors that impact the formation of latex film, including but not limited to temperature, pH, along with the incorporation of additives. The chapter delves into latex bonding mechanisms, outlining the concept of latex connecting and its relevance across diverse industries. This study delves into the interfacial phenomena and adhesion mechanisms integral to latex bonding, focusing on the physical interactions that underlie the strength of the bonding process. The present discourse centers on the mechanical characteristics of latex films, with particular emphasis on the strength of tensile along with elongation (Pradhan et al., 2023).

REVIEW QUESTIONS

1. What are the key factors that influence latex film formation mechanisms?
2. Describe the bonding mechanisms involved in latex bonding and adhesion.
3. How do the molecular structure and composition of latex particles affect their rheological behavior?
4. What are the mechanical properties of latex films, and what factors influence their tensile strength and elongation?
5. Explain the role of surface preparation techniques in enhancing latex bonding strength.
6. What are some potential future directions and emerging trends in latex technology and applications?

REFERENCES

1. Afiq, M. M., & Azura, A. R., (2013). Effect of sago starch loadings on soil decomposition of natural rubber latex (NRL) composite films mechanical properties. International Biodeterioration & Biodegradation, 85, 139–149.
2. Almeida, L. M., Floriano, J. F., Ribeiro, T. P., Magno, L. N., Da Mota, L. S. L. S., Peixoto, N., & Gonçalves, P. J., (2014). Hancornia speciosa latex for biomedical applications: Physical and chemical properties, biocompatibility assessment and angiogenic activity. Journal of Materials Science: Materials in Medicine, 25, 2153–2162.
3. Bluestein, C., Loewrigkeit, P., Mcgimpsey, T. T., & Van, D. K. A., (1975). Factors influencing latex tie-coat adhesion in urethane skin coat to fabric bonding. Journal of Coated Fabrics, 4(3), 158–165.
4. Cegala, D. J., (2006). Emerging trends and future directions in patient communication skills training. Health Communication, 20(2), 123–129.
5. Charmeau, J. Y., Berthet, R., Gringreau, C., Holl, Y., & Kientz, E., (1997). Effects of film structure on mechanical and adhesion properties of latex films. International Journal of Adhesion and Adhesives, 17(2), 169–176.
6. Dobler, F., & Holl, Y., (1996). Mechanisms of latex film formation. Trends in Polymer Science, 5(4), 145–151.
7. Dragnevski, K. I., & Donald, A. M., (2008). An environmental scanning electron microscopy examination of the film formation mechanism of novel acrylic latex. Colloids and Surfaces A: Physicochemical and Engineering Aspects, 317(1–3), 551–556.
8. Dralle, H., Lorenz, K., & Machens, A., (2012). Verdicts on malpractice claims after thyroid surgery: Emerging trends and future directions. Head & Neck, 34(11), 1591–1596.
9. Fathy, A., Zhu, H., & Kohail, M., (2022). Factors affecting the fresh-to-hardened concrete repair system. Construction and Building Materials, 320, 126279.
10. Gill, S. S., Xu, M., Ottaviani, C., Patros, P., Bahsoon, R., Shaghaghi, A., & Uhlig, S., (2022). AI for next generation computing: Emerging trends and future directions. Internet of Things, 19, 100514.
11. Goyal, S., Sergi, B. S., & Esposito, M., (2019). Literature review of emerging trends and future directions of e-commerce in global business landscape. World Review of Entrepreneurship, Management and Sustainable Development, 15(1, 2), 226–255.
12. Henson, W. A., Taber, D. A., & Bradford, E. B., (1953). Mechanism of film formation of latex paint. Industrial & Engineering Chemistry, 45(4), 735–739.

13. Hu, R., Dimonie, V. L., El - Aasser, M. S., Pearson, R. A., Hiltner, A., Mylonakis, S. G., & Sperling, L. H., (1997). Multicomponent latex IPN materials: 2. Damping and mechanical behavior. Journal of Polymer Science Part B: Polymer Physics, 35(10), 1501–1514.
14. Joanicot, M., Wong, K., & Cabane, B., (1996). Interdiffusion in cellular latex films. Macromolecules, 29(14), 4976–4984.
15. Keddie, J. L., (1997). Film formation of latex. Materials Science and Engineering: R: Reports, 21(3), 101–170.
16. Koyuncu, B., Hamori, M., & Baruch, Y., (2017). Guest editors' introduction: CEOs' careers: Emerging trends and future directions. Human Resource Management, 56(2), 195–203.
17. Kuppusamy, S., Thavamani, P., Venkateswarlu, K., Lee, Y. B., Naidu, R., & Megharaj, M., (2017). Remediation approaches for polycyclic aromatic hydrocarbons (PAHs) contaminated soils: Technological constraints, emerging trends, and future directions. Chemosphere, 168, 944–968.
18. Lee, D. I., (2005). The effects of latex coalescence and interfacial crosslinking on the mechanical properties of latex films. Polymer, 46(4), 1287–1293.
19. Lee, N. K., Kim, E. M., & Lee, H. K., (2016). Mechanical properties and setting characteristics of geopolymer mortar using styrene-butadiene (SB) latex. Construction and Building Materials, 113, 264–272.
20. Lepizzera, S., Lhommeau, C., Dilger, G., Pith, T., & Lambla, M., (1997). Film - forming ability and mechanical properties of coalesced latex blends. Journal of Polymer Science Part B: Polymer Physics, 35(13), 2093–2101.
21. Limousin, E., Ballard, N., & Asua, J. M., (2019). The influence of particle morphology on the structure and mechanical properties of films cast from hybrid latexes. Progress in Organic Coatings, 129, 69–76.
22. Ludlow, B. L., & Brannan, S. A., (1999). Distance education programs preparing personnel for rural areas: Current practices, emerging trends, and future directions. Rural Special Education Quarterly, 18(3, 4), 5–20.
23. Magbool, H. M., & Tayeh, B. A., (2021). Influence of substrate roughness and bonding agents on the bond strength between old and new concrete. Advances in Concrete Construction, 12(1), 33.
24. Manroshan, S., & Baharin, A., (2005). Effect of nanosized calcium carbonate on the mechanical properties of latex films. Journal of Applied Polymer Science, 96(5), 1550–1556.
25. Mazarji, M., Bayero, M. T., Minkina, T., Sushkova, S., Mandzhieva, S., Tereshchenko, A., & Keswani, C., (2021). Realizing united nations

sustainable development goals for greener remediation of heavy metals-contaminated soils by biochar: Emerging trends and future directions. Sustainability, 13(24), 13825.

26. McMahan, C., Cornish, K., Pawlowski, H., & Williams, J., (2005). Dynamic mechanical properties of latex films. In: Proceedings of the 167th Technical Meeting of the ACS Rubber Division. San Antonio.
27. Misra, S. C., Pichot, C., El - Aasser, M. S., & Vanderhoff, J. W., (1983). Batch and semicontinuous emulsion copolymerization of vinyl acetate–butyl acrylate. II. Morphological and mechanical properties of copolymer latex films. Journal of Polymer Science: Polymer Chemistry Edition, 21(8), 2383–2396.
28. Nalla, M. K., & Gurinskaya, A., (2020). Private police and security governance: Mapping emerging trends and future directions. Journal of Contemporary Criminal Justice, 36(1), 101–109.
29. Norhazariah, S., Azura, A. R., Sivakumar, R., & Azahari, B., (2016). Effect of different preparation methods on crosslink density and mechanical properties of carrageenan filled natural rubber (NR) latex films. Procedia Chemistry, 19, 986–992.
30. Ogbebor, J. O., Imoize, A. L., & Atayero, A. A. A., (2020). Energy efficient design techniques in next-generation wireless communication networks: Emerging trends and future directions. Wireless Communications and Mobile Computing, 2020, 1–19.
31. Ortega-Vinuesa, J. L., & Bastos-González, D., (2001). A review of factors affecting the performances of latex agglutination tests. Journal of Biomaterials Science, Polymer Edition, 12(4), 379–408.
32. Pan, Y., Wang, F., Wei, T., Zhang, C., & Xiao, H., (2016). Hydrophobic modification of bagasse cellulose fibers with cationic latex: Adsorption kinetics and mechanism. Chemical Engineering Journal, 302, 33–43.
33. Pan, Y., Zhang, M., Zhang, J., Zhu, X., Bian, H., & Wang, C., (2020). Effect of silane coupling agent on modification of areca fiber/natural latex. Materials, 13(21), 4896.
34. Poompradub, S., Thirakulrati, M., & Prasassarakich, P., (2014). In: Situ generated silica in natural rubber latex via the sol–gel technique and properties of the silica rubber composites. Materials Chemistry and Physics, 144(1, 2), 122–131.
35. Poovaneshvaran, S., Hasan, M. R. M., & Jaya, R. P., (2020). Impacts of recycled crumb rubber powder and natural rubber latex on the modified asphalt rheological behavior, bonding, and resistance to shear. Construction and Building Materials, 234, 117357.

36. Pradhan, V., Kumar, A., & Dhar, J., (2023). Emerging trends and future directions in software reliability growth modeling. Engineering Reliability and Risk Assessment, 131–144.
37. Quach, A., (1973). Polymer coatings. physics and mechanics of leveling. Industrial & Engineering Chemistry Product Research and Development, 12(2), 110–116.
38. Rith, M., Kim, Y. K., Lee, S. W., Park, J. Y., & Han, S. H., (2016). Analysis of in situ bond strength of bonded concrete overlay. Construction and Building Materials, 111, 111–118.
39. Roslim, R., Tan, K. S., & Jefri, J., (2018). Study on morphological structures and mechanical properties of natural rubber latex films prepared at different prevulcanization and drying temperatures. Journal of Rubber Research, 21, 1–16.
40. Routh, A. F., & Russel, W. B., (1999). A process model for latex film formation: Limiting regimes for individual driving forces. Langmuir, 15(22), 7762–7773.
41. Routh, A. F., & Russel, W. B., (2001). Deformation mechanisms during latex film formation: Experimental evidence. Industrial & Engineering Chemistry Research, 40(20), 4302–4308.
42. Santipanusopon, S., & Riyajan, S. A., (2009). Effect of field natural rubber latex with different ammonia contents and storage period on physical properties of latex concentrate, stability of skim latex and dipped film. Physics Procedia, 2(1), 127–134.
43. Selomulya, C., Bushell, G., Amal, R., & Waite, T. D., (2002). Aggregation mechanisms of latex of different particle sizes in a controlled shear environment. Langmuir, 18(6), 1974–1984.
44. Shaffie, E., Arshad, A. K., Alisibramulisi, A., Ahmad, J., Hashim, W., Abd, R. Z., & Jaya, R. P., (2018). Effect of mixing variables on physical properties of modified bitumen using natural rubber latex. International Journal of Civil Engineering and Technology, 9(7), 1812–1821.
45. Shahani, A., Shiffler, D. A., Batra, S. K., & Cannon, C. A., (1999). Foamed latex bonding of spun lace fabrics to improve physical properties. International Nonwovens Journal, (2), 1558925099OS-800217.
46. Shankar, V., & Hollinger, M., (2007). Online and mobile advertising: Current scenario, emerging trends, and future directions. Marketing Science Institute, 31(3), 206–207.
47. Sheetz, D. P., (1965). Formation of films by drying of latex. Journal of Applied Polymer Science, 9(11), 3759–3773.

48. Shunmuga, V. P. R., & Ammasi, A. K., (2023). A comprehensive review on the factors affecting bond strength in concrete. Buildings, 13(3), 577.
49. Stevens, M. J., Steren, R. E., Hlady, V., & Stewart, R. J., (2007). Multiscale structure of the underwater adhesive of Phragmatopoma californica: A nanostructured latex with a steep microporosity gradient. Langmuir, 23(9), 5045–5049.
50. Steward, P. A., Hearn, J., & Wilkinson, M. C., (2000). An overview of polymer latex film formation and properties. Advances in Colloid and Interface Science, 86(3), 195–267.
51. Teixeira, R. F., & Bon, S. A., (2010). Physical methods for the preparation of hybrid nanocomposite polymer latex particles. Hybrid Latex Particles: Preparation with (Mini) Emulsion Polymerization, 19–52.
52. Tian, M., Qi, Y., & Oh, E. S., (2021). Application of a polyacrylate latex to a lithium iron phosphate cathode as a binder material. Energies, 14(7), 1902.
53. Vanderhoff, J. W., (1970). Mechanism of film formation of latices. British Polymer Journal, 2(3), 161–173.
54. Vansteenkiste, M., & Mouratidis, A., (2016). Emerging trends and future directions for the field of motivation psychology: A special issue in honor of Prof. Dr. Willy Lens. Psychologica Belgica, 56(3), 317.
55. Vansteenkiste, M., Niemiec, C. P., & Soenens, B., (2010). The development of the five mini-theories of self-determination theory: An historical overview, emerging trends, and future directions. The Decade Ahead: Theoretical Perspectives on Motivation and Achievement, 16, 105–165.
56. Wang, J., Wen, Y., Ge, H., Sun, Z., Zheng, Y., Song, Y., & Jiang, L., (2006). Simple fabrication of full color colloidal crystal films with tough mechanical strength. Macromolecular Chemistry and Physics, 207(6), 596–604.
57. Wang, L., Loh, K. J., Brely, L., Bosia, F., & Pugno, N. M., (2016). An experimental and numerical study on the mechanical properties of carbon nanotube-latex thin films. Journal of the European Ceramic Society, 36(9), 2255–2262.
58. Wang, L., Tian, Z., Ma, G., & Zhang, M., (2020). Interlayer bonding improvement of 3D printed concrete with polymer modified mortar: Experiments and molecular dynamics studies. Cement and Concrete Composites, 110, 103571.
59. Wicks, D. A., Rhoades, A. M., Williamson, J. S., & Miriyala, B., (2006). Investigation of antimicrobial peptide salts: Efficacy and solubility in surfactant solutions for latex systems. In: Combinatorial Methods and Informatics in Materials Science (pp. 139–146).

60. Winnik, M. A., (1997). Latex film formation. Current Opinion in Colloid & Interface Science, 2(2), 192–199.

61. Xie, H. Q., Liu, H., Liu, Z. H., & Guo, J. S., (1996). Electrically conductive polyaniline/poly (butadiene - co - styrene - co - 2 - vinylpyridine) latex composites. Die Angewandte Makromolekulare Chemie: Applied Macromolecular Chemistry and Physics, 243(1), 117–128.

62. Xu, F., Peng, C., Zhu, J., & Chen, J., (2016). Design and evaluation of polyester fiber and SBR latex compound-modified perlite mortar with rubber powder. Construction and Building Materials, 127, 751–761.

63. Xu, K., Fan, B., Putera, K., Wawryk, M., Wan, J., Peng, B., & Thang, S. H., (2022). Nanoparticle surface cross-linking: A universal strategy to enhance the mechanical properties of latex films. Macromolecules, 55(13), 5301–5313.

64. Xu, L. C., & Logan, B. E., (2006). Adhesion forces between functionalized latex microspheres and protein-coated surfaces evaluated using colloid probe atomic force microscopy. Colloids and Surfaces B: Biointerfaces, 48(1), 84–94.

65. Zhang, L., Melechko, A. V., Merkulov, V. I., Guillorn, M. A., Simpson, M. L., Lowndes, D. H., & Doktycz, M. J., (2002). Controlled transport of latex beads through vertically aligned carbon nanofiber membranes. Applied Physics Letters, 81(1), 135–137.

66. Zhong, S., & Chen, Z., (2002). Properties of latex blends and its modified cement mortars. Cement and Concrete Research, 32(10), 1515–1524.

67. Zosel, A., & Ley, G., (1993). Influence of crosslinking on structure, mechanical properties, and strength of latex films. Macromolecules, 26(9), 2222–2227.

68. Zosel, A., (1995). Mechanical properties of films from polymer latices. Polymers for Advanced Technologies, 6(5), 263–269.

CHAPTER

4 Latex Preparation and Characterization

Contents

4.1. INTRODUCTION

This chapter centers on preparing, formulating, and characterizing latex materials. Latex is a colloid type consisting of polymer particles dispersed in water (Xiong et al., 2003). The characteristics and efficacy of latex are subject to the influence of multiple factors. Comprehending the techniques involved in the preparation, considerations in formulation, and methods for characterization are crucial in attaining the desired properties of latex and enhancing their performance in application. The initial section of the chapter delves into the process of latex preparation, encompassing various facets such as the formulation of recipes and the implementation of feed strategies. The recipe formulation process entails carefully selecting suitable monomers, additives, and initiators to attain the intended polymer structure and functionality (Covolan et al., 1997).

In contrast, feed strategies pertain to the regulation of monomer and initiator introduction throughout the polymerization procedure to enhance the quality of the latex. The utilization of sensors and process control is of paramount importance in the production of latex. The chapter examines using sensors and monitoring systems to ensure adequate regulation of critical parameters, including temperature, pH, and pressure (Xiong et al., 2002). The implementation of control systems aids in the preservation of uniformity and replicability in the production of latex, thereby resulting in a sustained level of product excellence. The preparation of latex is contingent upon crucial factors such as reactor agitation and temperature. Adequate agitation is crucial for achieving optimal reactant mixing and promoting polymerization, whereas temperature regulation plays a significant role in determining reaction kinetics and the resultant properties of the latex. This chapter analyzes the influence of reactor agitation and temperature on latex formation and explores techniques for enhancing these variables (Wu & Chen, 2008).

The assessment of the properties and performance of latex is contingent upon its characterization. The chapter encompasses a range of characterization techniques, including the assessment of colloidal stability. This technique serves to gauge the stability of latex dispersions over some time. The polymerization process can be evaluated through conversion measurement, quantifying the degree of monomer conversion into the polymer. This analysis can offer valuable information regarding the efficiency of the reaction (Noguchi et al., 1993). Understanding the properties of latex necessitates a comprehensive understanding of critical parameters such as copolymer composition, molecular weight, size distribution, particle size, and polymerization rate. The present chapter delves into the techniques employed for characterizing the parameters mentioned above, emphasizing their impact on the performance of latex and its suitability for various applications. The rheology of latex is analyzed to comprehend the flow and stretching characteristics of latex substances. Materials' rheological characteristics are paramount in their processing and application performance. This chapter delves into the various rheological measurement techniques utilized for evaluating latex behavior and examines their implications for film formation and mechanical properties. The film formation process holds significant importance in the context of latex applications. This chapter analyzes the various factors that impact the formation of films in the context of latex, including the rheology of the substance, the characteristics of particle surfaces, and the conditions under which drying occurs. Comprehending the film formation process is crucial in optimizing coating, adherence, and film characteristics for diverse applications (Cui et al., 2008). The performance of latex films is contingent upon their mechanical properties. The present chapter explores various techniques for characterizing the mechanical properties of latex films, encompassing tensile strength, elongation, and modulus. The abovementioned characteristics are pivotal in ascertaining the film's resilience, flexibility, and comprehensive efficacy. The chapter ultimately delves into techniques for characterizing the surface of particles, specifically those of latex, to better understand their surface chemistry, functional groups, and interactions. Characterizing surface properties is of utmost importance in comprehending the stability of latex, its adhesion, and how it interacts with other materials. This chapter generally overviews the diverse facets of planning, formulating, and characterizing latex. The aforementioned emphasizes the significance of comprehending said factors in attaining the intended characteristics of latex and enhancing their efficacy in diverse applications (Alex & Nah, 2006).

4.2. LATEX PREPARATION

4.2.1. Recipe Formulation

The chemical composition, also known as the formulation, is a crucial determinant of the form and characteristics of the end product. Certain constituents are incorporated into the formulation either prior to or during the process of polymerization. Post-additives are incorporated after the polymerization process to alter the properties of the latex. Emulsions can be divided into two distinct phases: the dispersed phase, which is composed of oil and contains the monomers and other components that are soluble in the monomer, and the continuous phase, which is composed of water and contains components that are soluble in water (Bao et al., 2015). The present discourse delineates the distinct constituents that constitute the oil and aqueous phases in an emulsion, along with their respective functionalities.

4.2.1.1. Polymerization Components

Typically, these constituents are amalgamated before undergoing polymerization to generate an emulsion. In specific instances, certain components may be incorporated during the polymerization process.

Monomers and Comonomers. The selection of monomers is a crucial aspect of emulsion polymerization, wherein the performance specifications of the target application serve as the primary basis for their choice (Liao et al., 2023). The consideration of cost is a crucial aspect when choosing a suitable monomer. The monomer's miscibility with the water phase cannot be absolute, as this would result in dispersion polymerization. Similarly, complete insolubility would impede conventional emulsion polymerization. Most monomers exhibit limited solubility in aqueous solutions and conform to these parameters.

The properties of latexes derived from homopolymers are predetermined due to their production from a single monomeric species. Any modification of these properties would require the inclusion of supplementary chemical components. Due to this rationale, most commercial emulsions comprise a comonomer system, which is a blend of various monomers instead of a solitary monomer. The copolymer's characteristics, such as its glass transition temperature, mechanical properties, and solvent resistance, can be regulated by selecting a suitable comonomer system (Burfield & Ng, 1978).

Typically, monomers are supplemented with inhibitors (e.g., 4-tert-butylcatechol) to impede untimely thermal polymerization during storage or transit. In laboratory experiments that require precise control, the removal of inhibitors is achieved through distillation. Nonetheless, inhibitor removal is not a prerequisite for industrial processes. Supplementary quantities of initiator are incorporated to counterbalance the existence of inhibitor.

Crosslinkers: The enhancement of latexes' chemical and mechanical characteristics can be achieved by integrating crosslinking monomers within the oil phase of the emulsion formulation. Monomers that contain multiple reactive groups, such as divinyl benzene or ethylene glycol dimethacrylate, can create crosslinked polymer networks within particles, resulting in the formation of microgels (homogeneous crosslinking). These crosslinked networks can also be formed at the interfaces between adjacent latex particles (interfacial crosslinking) or interstices between the particles (interstitial crosslinking) (Liu et al., 1999). The crosslinks in a given system may manifest as covalent, ionic, or hydrogen-bonding interactions. The concentration of the crosslinking agent is a determinant factor in regulating the crosslink density (mc). Grafting reactions, which involve gel formation in polyvinyl acetate polymerizations by chain transfer, are a viable means by which crosslinking can occur in emulsion polymerization.

Chain transfer agents are compounds that can interrupt the growth of a polymer chain by transferring a radical to another molecule. These agents are commonly used in chain transfer agents, such as t-butyl mercaptan, which are typically incorporated into the oil phase of latex production. These agents function to abstract radicals from the polymer chains as they grow, resulting in a reduction of the average molecular weight of the polymer within the latex particles (Dubois et al., 1994). The addition of chain transfer agents is a common practice in polymerization processes aimed at regulating the ultimate molecular weight of the polymer or constraining the untimely emergence of gel.

Initiators: The initiator plays a crucial role in emulsion polymerization by serving as a nucleating agent for the particles and facilitating a consistent flow of radicals throughout the polymerization process. Ionic initiators utilized in emulsion polymerization, such as ammonium persulfate, tend to be water-soluble. Latexes initiated by persulfate exhibit an increased anionic surface charge, which facilitates the maintenance of colloidal stability (Schneider et al., 2002). Additional initiators, such as hydroperoxides (e.g., cumene hydroperoxide), exhibit a nonionic nature and varying degrees of water solubility. Certain initiators, namely benzoyl peroxide (BPO) and 2,2'-azobis(isobutyronitrile) (AIBN), exhibit exclusive solubility in oil. Redox initiator systems, such as the combination of tert-butyl hydroperoxide and sodium formaldehyde sulfoxylate, utilize reduction-oxidation chemistry to produce radicals at reduced temperatures, thereby enabling enhanced regulation of reactor temperature and facilitating high molecular weight advancement. Many novel initiators are currently under development to cater to specific applications. The selection of the initiator is primarily based on its partitioning behavior between the oil and aqueous phases, as well as its half-life (Vanderhoff, 1993).

The selection of a suitable surfactant (i.e., stabilizer) for a specific emulsion polymerization process is determined by various factors. The surfactant plays a crucial role in regulating the size of latex particles and ensuring their colloidal stability. An excessive amount of coagulum present in the product could potentially suggest inadequate colloidal stability. Augmenting the quantity of surfactant in a formulation can enhance its stability. However, exceeding the optimal amount may trigger the formation of undesired particles through secondary nucleation. Surfactants possessing hydrophilic/lipophilic balance (HLB) values greater than 7 are employed in formulating conventional oil-in-water emulsions (Boutti et al., 2005).

In contrast, surfactants with HLB values lower than 7 are utilized to prepare inverse water-in-oil emulsions. It is imperative to consider the segregation of the surfactant into either the oil or aqueous phases. Minimizing the quantity of surfactant employed in a formulation is commonly favored due to economic considerations and concerns related to the characteristics and efficacy of the product. The abundance of surfactants can influence the water sensitivity of films derived from latexes (Guyot et al., 2007). A diverse range of surfactants can be obtained commercially.

Emulsions have been prepared using anionic surfactants. Diverse surfactants in a given formulation can confer supplementary stability in particular circumstances. Mixtures comprising anionic and nonionic surfactants are frequently employed. The anionic surfactant regulates the particle nucleation stage, while the nonionic surfactant confers supplementary electrolyte tolerance, mechanical shear stability, and freeze-thaw stability (Ceska, 1974). It is advisable to steer clear of blends containing both anionic and cationic surfactants due to their propensity to coagulate.

Certain surfactants exhibit multifunctionality. The chemical reaction between monomer and reactive surfactants, also known as surfmers, results in their binding at the surface of particles. This binding is irreversible and prevents their desorption or migration to the surface of a latex film. Previous studies reported these findings. Surfactants may possess reactive groups that serve as initiators, known as inisurfs, or chain transfer agents, referred to as transurfs. Defoamers, also known as anti-foaming agents, are commonly utilized in emulsion formulations to mitigate the foaming properties of surfactants. These agents are typically composed of silicones or hydrocarbons (Asua, 2004).

Buffers play a crucial role in maintaining the pH of the aqueous phase during the decomposition of persulfate initiators, which leads to the formation of radicals. The degradation rate of persulfate initiators is accelerated with a decrease in the pH of the aqueous phase, resulting in untimely exhaustion of the initiator. Aqueous phase buffers, such as sodium bicarbonate, regulate the pH and prevent significant fluctuations. The electrolytic nature of buffers can impact the colloidal stability of latex particles, potentially leading to alterations in particle size. Water is a polar solvent that constitutes the majority of the aqueous phase and is a ubiquitous component in emulsion polymerizations. The solubility of organic liquids, such as monomers, in water, is restricted due to their non-polar nature. This facilitates the dissemination of distinct monomer particles. The aqueous phase comprises an initiator, surfactant, and buffer. Due to its high heat capacity, water can extract significant amounts of thermal energy during polymerization. The low density of the water medium facilitates rapid mixing and heat transfer. Water removal can be facilitated through the coagulation of latex particles or via evaporation (Dimitratos et al., 1994). Water is a cost-effective resource that does not adversely affect human health or the environment. The presence of impurities, such as dissolved salts and electrolytes, may impact the polymerization process and the ultimate properties of the latex. Moreover, the transportation cost of water in an industrial context is considerably high. As a consequence, there has been a development of high solids latexes.

4.2.1.2. Post-polymerization Additives

Upon the completion of polymerization, it may be necessary to incorporate certain additives to augment the functionality or stability characteristics of the resultant latex product. It could be deemed imperative to alter the viscosity of the latex, enhance its film-forming capacity, or safeguard it against biological or oxidative deterioration (Issart et al., 2019) (Figure 4.1).

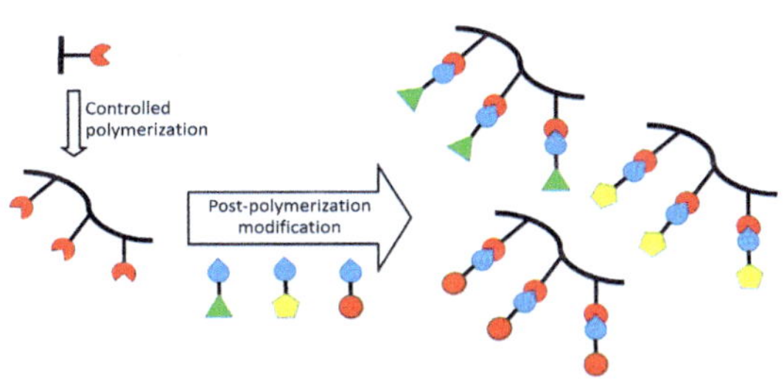

Figure 4.1. Schematic illustration of the post polymerization modifications.

Source: Xu Chen, Creative Commons License.

Surfactants, or post-stabilizers, are chemical compounds commonly used in various industries. Latexes must exhibit stability in the absence of shear at ambient temperature. The latex particles may undergo shear-induced coagulation when subjected to shear forces during the transfer process from the reactor to a tank or from tank to tank. Latexes are frequently supplemented with an extra surfactant, which is referred to as a post-stabilizer, to enhance their shear stability. In regions with low temperatures, latexes may encounter severe weather conditions that initiate the freezing of the aqueous phase (Piedra-Cascón et al., 2021). This process leads to the compression of latex particles and consequent coagulation, as stated by the source. Additional stabilizers are incorporated to enhance the product's freeze-thaw stability and electrolyte tolerance. Latexes are commonly characterized by low viscosity owing to their aqueous dispersion nature, necessitating rheology modifiers (Hayashi et al., 1964). Nevertheless, it might be imperative to augment the viscosity of the ultimate latex outcome under certain circumstances. The appropriate viscosity of latex paints is crucial to ensure uniform flow on a substrate while preventing any undesired dripping. The latex viscosities can be manipulated by incorporating a viscosity modifier, commonly referred to as a thickener, into the aqueous phase. The process of coalescing solvents involves the merging of two or more solvents into a single phase. Coalescing aids, specifically ester alcohols, can serve as plasticizers by reducing the glass transition temperature of latex. This results in an increased rate of polymer chain interdiffusion between neighboring particles during film formation (Espeel et al., 2023). The coalescent agents' volatility should be sufficiently low to enable water evaporation, followed by film formation in the coalescent agent's presence. Nonetheless, the coalescing agent must evaporate at a sufficiently rapid rate to enable the film to solidify after its formation, thereby elevating the glass transition temperature to that of the latex material in isolation.

Polymers are susceptible to degradation over time due to free-radical oxidation. Antioxidants and UV stabilizers are commonly employed to mitigate this process. Polymers are fortified with antioxidants, including hindered phenols, which function as free radical scavengers to impede degradation. The autocatalytic nature of the oxidation process implies that the degradation rate escalates in proportion to the concentration of free radicals (Mülhaupt, 2003). Polymers are subject to various sources of free radical formation, such as residual initiator, stress-induced radical formation, high temperature, and ultraviolet (UV) light exposure. UV stabilizers, such as 2-hydroxy benzophenone, absorb ultraviolet radiation, thereby safeguarding the polymer structure from damage. The topic of interest pertains to biocides and fungicides. Latex coatings, such as those used in paints, are frequently subjected to outdoor conditions, rendering them susceptible to microbial and fungal proliferation. Indoor settings, such as bathrooms and kitchens, also harbor high moisture conditions that facilitate the proliferation of microorganisms. Incorporating biocides and fungicides into latex is a viable method for preventing microbiological attacks. Metallic and halogenated compounds are commonly found in biocide reagents.

4.2.2. Feed Strategies

The dimensions, configuration, and constitution of latex particles are contingent not solely upon the constituents employed in the emulsion formulation but also on the modality of amalgamation and polymerization of said constituents. By regulating the feed strategy, it is possible to alter the emulsion polymerization process and generate latex particles that possess distinct characteristics (Li et al., 2015). The selection of an optimal feed strategy is contingent upon individual circumstances, as each strategy presents its own set of advantages and disadvantages. The primary feeding techniques are enumerated below and are visually depicted in Figure 4.2.

Tip

Remember to carefully consider the feed strategies used in latex preparation as they can significantly impact the quality and properties of the final product.

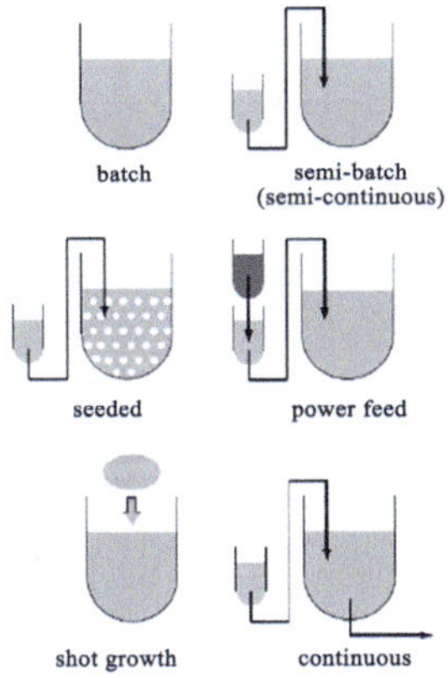

Figure 4.2. Emulsion polymerization feed strategies shot growth continuously.

Source: Daniels, Creative Commons License.

4.2.2.1. Batch

The batch polymerization process involves introducing the emulsion components into a reactor equipped with a stirring mechanism, followed by applying heat to initiate the polymerization reaction. There is no introduction or elimination of any substances throughout the polymerization process (Guo et al., 2014). Due to the highly exothermic nature of most polymerizations, there is a risk that the heat generation rate may surpass the heat removal capacity, leading to the potential occurrence of a runaway reaction. The batch polymerization technique provides limited or negligible regulation over the copolymer composition, relying instead on the comonomer reactivity ratios and the division of the comonomers in the latex particles.

The batch process is a straightforward approach that applies to various polymerization reactions. However, the control over the reaction is limited to the recipe, temperature, and agitation conditions. The mono-dispersity of latex particles is typically achieved through nucleation during the initial stages of batch polymerization (Asua, 2004). This process has been utilized to produce polystyrene-size standard particles. Nevertheless, notable discrepancies may persist among batches, posing challenges to achieving uniformity in product manufacturing through the batch method.

Batch polymerizations are frequently conducted during screening experiments at the laboratory scale. Batch polymerizations are not as commonly employed in commercial production processes at a large scale as semi-continuous polymerizations due to the inherent constraints in copolymer composition control and heat transfer (Flores-Cerrillo & MacGregor, 2002).

4.2.2.2. Semi-Continuous

The semi-continuous method, also called semi-batch, is the preferred approach for conducting most emulsion polymerization procedures. Semi-continuous polymerizations are comprised of two distinct stages, namely the planting stage and the feeding stage. In the seeding phase, a proportion of the complete emulsion is introduced into a reactor, followed by the addition of an initiator to initiate polymerization (Jönsson et al., 2007). The seeding stage establishes the total count of seed particles, which should remain consistent throughout the feeding stage, notwithstanding the growth in particle size.

During the feeding stage, an extra monomer is continuously supplied to the reactor to provide monomer to the polymerizing particles. This is often done in the form of an emulsion to enhance mass transfer and minimize the likelihood of monomer pooling. The regulation of the monomer addition rate enables the regulation of the heat generation rate. Adding an initiator during the polymerization process can regulate the radical flux, polymerization rate, and heat generation rate (Sosa et al., 2001). The addition of surfactant during the reaction process can serve to stabilize the polymer particles as they grow. It is imperative to avoid excessive pumping of surfactant during polymerization, as this may lead to the formation of unwanted particles through secondary nucleation.

The copolymer composition can be regulated by the corresponding rates of monomer addition in copolymerizations. By utilizing this approach, it is possible to mitigate significant variations in comonomer reactivity ratios or water solubilities, thereby yielding a copolymer that exhibits consistent composition (Xu et al., 2013). To regulate monomer concentration within polymer particles, it may be necessary to conduct the polymerization under conditions of limited monomer availability. The statement mentioned above implies that the polymer particles are undergoing polymerization with an instantaneous conversion rate of 90% or higher, indicating that they are not fully saturated with monomers. Suppose the rate of monomer addition surpasses the rate of polymerization. In that case, the reactor will be deemed operating under flooded conditions, resulting in a loss of control over the copolymer composition.

The copolymer composition is frequently regulated to modulate the glass transition temperature (T_g) of the copolymer (Vanderhoff, 1987). An instance of this could be altering the proportion of vinyl acetate to butyl acrylate to generate a copolymer with a glass transition temperature near ambient temperature, intended for utilization in latex paint applications. In certain instances, the water-resistant properties of the polymer are enhanced through the copolymerization of hydrophobic monomers. An example of this is the copolymerization of styrene with butadiene, which is utilized for nonwoven textile applications (Zhang et al., 2015).

4.2.2.3. Shot-Growth

Occasionally, it becomes imperative to introduce a bulk amount of substance into a reactor currently undergoing emulsion polymerization. To produce a functionalized particle surface, a minor quantity of functional monomer may be incorporated into a latex towards the conclusion of the polymerization process (Chainey et al., 1982). The procedure above is commonly called the shot-growth technique. The technique above is frequently employed in emulsifier-free polymerizations, which lack surfactants but instead utilize functional monomers that are highly water-soluble and surface-active. In the initial phases of emulsifier-free polymerization, the size of the particles is predominantly influenced by the initiator, for instance, potassium persulfate, and the functional monomer, such as sodium styrene sulfonate. Latexes produced using this method tend to have a narrow size distribution. When the conversion rate reaches approximately 90%, an extra functional monomer dose is introduced. This functional monomer chemically adheres to the surface of the particles, thereby modifying the surface charge density. Industrial coatings can be crosslinked with functional monomers such as acrylic acid, which offer carboxyl sites (Twigt, et al., 1993).

4.2.2.4. Seeded

Seed latexes are frequently utilized in producing numerous product batches that require a precise and uniform particle size. Seeded emulsion polymerizations, as reported in the literature, entail swelling a typical batch of seed latex particles with monomer, followed by polymerization and growth to a predetermined size. The addition of surfactant is necessary to ensure the stabilization of the surface area of the growing particle (Plessis et al., 2001). However, it is imperative to avoid excessive surfactant usage as it may result in secondary nucleation. The polymerization process does not generate any additional particles, thereby establishing the particle count based on the initial number present in the well-defined seed latex. Seeded emulsion polymerizations are a viable method for precisely controlling the growth of particles to a predetermined final size. The distribution of particles resulting from seeded polymerizations exhibits a high degree of uniformity. Core-shell nanoparticles can be synthesized by introducing a comonomer that differs from the one utilized in the seed particles. The challenge associated with employing this technique is that upon depletion of the initial latex seed, generating an additional set of particles possessing identical dimensions becomes necessary. Alternatively, it is necessary to modify the seeded polymerization process in order to accommodate the variations in the seed latexes.

4.2.2.5. Power Feed

The power feed technique is a semi-continuous process that is employed to modulate the comonomer concentration during polymerization, thereby generating particles exhibiting a copolymer-gradient core-shell morphology

(Hawkett et al., 1980). In a binary comonomer system, utilizing two feed tanks and two constant-rate pumps is imperative. Rather than directly supplying each pump to the reactor, a method involves introducing one comonomer into the tank containing the other comonomer. This tank is thoroughly mixed, and both comonomers are then concurrently supplied to the reactor. The comonomer composition undergoes a continuous drift from the initial pure component at the onset of the reaction, traversing a fluctuating comonomer blend, culminating in a terminal comonomer composition. Acrylic copolymers with a core-shell gradient structure have been synthesized to serve as impact modifiers (Chern, 2006).

4.2.2.6. Continuous

The continuous emulsion polymerization process involves a steady supply of monomer emulsion being introduced into a reactor where polymerization occurs. The resulting product is then continuously extracted from the reactor (Poehlein & Dougherty, 1977). Continuous polymerizations are distinct from batch reactions in that they are conducted under steady-state conditions, ensuring that the end result remains consistent across all batches. In contrast, batch reactions may yield products that vary significantly from batch to batch. Furthermore, continuous processes exhibit higher production rates and offer improved control over heat removal compared to batch polymerizations (Schork & Ray, 1987).

Continuous reactors are typically classified into two main types: continuously stirred tank reactors (CSTRs) and plug flow reactors (PFRs). CSTRs are voluminous containers designed to achieve optimal mixing conditions, ensuring the emulsion composition is homogenous throughout the entire reactor volume during the polymerization process (Pauer, 2018). CSTRs are typically operated under isothermal conditions with a consistent overall conversion rate. CSTRs are frequently employed in a sequential or cascading manner to increase the degree of conversion gradually. The production of styrene-butadiene rubber has been carried out using this method. A disparity exists in the polymerization duration of latex particles within a CSTR. A heterogeneous mixture of particles exhibits a range of particle residence times, diameters, and compositions, resulting in the early exit of certain particles relative to others.

As opposed to CSTRs, PFRs exhibit a tubular configuration. The polymerization process commences as the monomer emulsion is introduced into the inlet of the plug flow reactor (PFR) (Nomura et al., 2005). The conversion level is directly proportional to the distance the emulsion covers during its passage through the Plug Flow Reactor (PFR). The ultimate latex is discharged from the terminal point of the tubular plug flow reactor (PFR). In an ideal scenario, every fluid element traversing a plug flow reactor (PFR) experiences uniform travel distance, velocity, and residence time. Consequently, the particles generated in a Plug Flow Reactor (PFR) exhibit a higher degree of uniformity in size and composition. Tubular reactors exhibit limited flexibility concerning analyzing and manufacturing capabilities in practical applications, and therefore, their utilization in emulsion polymerization processes is not widespread in industrial settings (Rawlings & Ray, 1988).

4.2.3. Sensors and Process Control

Due to the inherent sensitivity of emulsion polymerization, it is imperative to exercise caution to maintain the uniformity of the end product across multiple batches or within a single continuous operation. In addition, it is noteworthy that emulsion polymers are classified as "products by process," indicating that the methodology employed during polymerization may have a greater impact on the ultimate product form than the constituent materials themselves.

It is preferable to conduct online monitoring of the emulsion polymerization process. However, it is common practice for numerous commercial producers to measure properties after polymerization and subsequently modify the process.

Implementing efficient process control measures is of utmost importance in ensuring the safe and consistent operation of the production process, as evidenced by scholarly literature.

The parameters that are measured or inputted are obtained via sensors, subjected to analysis, refinement, or amplification, and subsequently transmitted to actuators to regulate the response or output parameters. The most basic form of controllers is feedback controllers, which receive input signals from sensors and generate output signals that are proportional to the input signals to drive the actuators (Gagnon & MacGregor, 1991). Emulsion polymerizations have been controlled using computer control methods. The strategies mentioned above have been derived from mathematical models of the emulsion polymerization process. The utilization of statistical process control is increasingly prevalent in the domain of emulsion polymerization.

4.2.4. Reactor Agitation

The agitator category encompasses a variety of types, including but not limited to pitched-blade impellers, baffled impellers, and the Rushton turbine. These agitators exhibit different levels of shear and mixing (Gonzaga et al., 2009). Modifying the agitation (stirring) level within a reactor will result in ramifications for the latex. Agitation comprises two fundamental constituents: shear, which refers to the tangential distortion of a small fluid element, and mixing, which denotes the comprehensive intermingling of the fluid. The principal consequences of shear involve the generation of vortices, disseminating of small liquid particles, and the emergence of coagulation induced by shear. The principal consequences of mixing pertain to the diminution of the stagnant volume, concomitant accumulation of monomer, and alterations in the coefficients of mass and heat transfer. The non-linear scaling of reactors with size can often obscure scale-up challenges, (Johnson, 1980). The heat transfer constraints in industrial-sized reactors are often attributed to the volume-to-surface area ratio increasing as the reactor size increases. Furthermore, during upscaling from a smaller reactor, it is customary to maintain a constant stirring power input per unit volume to ensure consistent mixing. As a result, the shear rate experiences an increment, leading to alterations in the reactor conditions.

4.2.5. Temperature

The optimal temperature for the reactor is determined by a delicate equilibrium between the duration of heating, the rate constant of propagation, the dissociation constant of the initiator, and the reactor's capacity for heat removal (Erdoğan et al., 2002). A significant release of heat typically characterizes the process of polymerization, and the polymerization rate is governed by an Arrhenius-type function in relation to temperature. It is imperative to avoid runaway (auto-acceleration) reactions due to the following factors. Upon initiation of polymerization, the need for external heating may be diminished.

In industrial-scale reactors with larger volumes ranging from 1,000 to 100,000 liters, the ratio of reactor volume to surface area is relatively high, leading to suboptimal heat dissipation (Vallittu et al., 1998). The utilization of cooling jackets is imperative in chemical processes, as it enables the maintenance of a constant temperature within the reactor or jacket during the course of the reaction. In certain polymerization processes, heat dissipation is negligible. The emulsion formulation is structured to prevent the adiabatic temperature increase within the reactor during polymerization from surpassing the maximum safe temperature threshold. This threshold is typically situated below the boiling point of the aqueous continuous phase. The prevailing conditions during emulsion polymerization are characterized by low pressures, which are in close proximity to atmospheric pressure (Xia et al., 1998). To eliminate the presence of atmospheric oxygen in the headspace of a reactor, a nitrogen purge or blanket is implemented to scavenge free radicals effectively. Otherwise, the occurrence of an induction period, which is a phase during which the polymerization process is inhibited, or a retardation effect, which is a decrease in the polymerization rate, may be noted. It should be noted that emulsion polymerizations can occur at pressures other than low pressure. Typically, high-pressure conditions are employed to polymerize vinyl acetate-ethylene (VAE), vinyl chloride, and styrene-butadiene rubber.

4.3. LATEX COLLOIDAL STABILITY CHARACTERIZATION

Colloidal instability in latexes may pose a problem if the latex is sheared during mixing or pumping; if the latex is frozen and then thawed; if contaminants or additives are added to the latex; or if the latex is stored for a long time at different temperatures (Daronch et al., 2006). The particles may be redispersed by sonication in situations when they have reversibly aggregated. By filtering the latex through a fine mesh sieve, the degree of coagulation can be calculated regarding the percentages of the total solids. In tests for shelf-life stability, the monomer droplets rise, and the polymer particles fall depending on the density of the dispersed phase compared to the continuous (aqueous) phase (Sansatsadeekul et al., 2011). Although particles may accumulate in a vessel over time to create concentrated bands (striations), if the surfactant successfully preserves the colloidal stability, the particles won't permanently clump together and can be redispersed. Similarly, if the surfactant is effective, droplets creamed at the top of a vessel will only very slowly assemble over time.

The creaming or settling rate can be seen by pouring a uniform emulsion or latex in a graduated cylinder and monitoring the movement of the front separating the opaque and transparent phases over time. Centrifugation can hasten the process of creaming or settling (Peula et al., 1994). Creaming/settling rate (also called sedimentation velocity, v) can be determined from droplet or particle diameter using the Stokes equation:

$$v = \frac{d^2(\rho - \rho_o)g}{18\eta}$$

where d denotes the size of the particles, denotes their density, o denotes the density of the continuous phase, g denotes the gravity acceleration constant, and denotes their viscosity.

4.4. CONVERSION TEST

In order to comply with ever-stricter environmental requirements, it is crucial to obtain a high monomer-to-polymer conversion rate in the majority of polymerizations (Amai et al., 1989). Additionally, recycling monomers numerous times before creating the final polymer product is rarely more effective than polymerizing monomers the first time through the reactor. In some circumstances, it could be challenging to achieve high conversion because the polymeric radicals can't penetrate highly viscous polymers to reach sparingly distributed molecules of unreacted monomer. For environmental or physical property reasons, post-processing could be required to reduce the product's residual monomer content. On the other hand, in some circumstances, low conversion may be required to avoid the creation of polymer gel networks (Calheiros et al., 2014). One such instance is in styrene-butadiene polymerizations, which are stopped at a conversion level of roughly 70% and then stripped of the monomer.

Gravimetric analysis is the most practical method for determining conversion in a reaction. Weighing a sample of partially polymerized latex is followed by evaporating the remaining water and monomer and weighing the dried polymer sample (Bouschlicher et al., 2000). The conversion can be computed using the original recipe, considering any ingredients added during the polymerization. Since a polymer's density typically exceeds its monomer's, conversion can be determined throughout the polymerization process by tracking a latex's change in density (also known as densitometry). Similarly, to this, it is possible to measure the volume change during polymerization (i.e., dilatometry to calculate the conversion. Gas chromatography (GC) can also be used to get conversion. Monomer will elute through the GC columns even though polymer won't, and conversion can be determined by integrating the response signal calibrated using an internal standard (Calheiros et al., 2008; Palin et al., 2014).

4.5. RATE OF POLYMERIZATION

The rates of polymerization, also known as polymerization kinetics, are of great importance because they directly impact the final product's mechanical, chemical, and physical characteristics. The polymerization rate increases in industrial applications to achieve the quickest cycle time and highest product production (Calheiros et al., 2006).

Data on conversion (x) against time (t) collected from densitometry and dilatometry have been used to estimate polymerization rates. The derivative (dx/dt) can be used to estimate the immediate polymerization rate. However, the scatter in the data and the quantity of conversion versus time data points limit the method's accuracy.

Reaction calorimetry is a more precise technique for measuring the reaction rate directly (Schneider et al., 2008). The rate of emulsion polymerization can be determined using any of a number of commercial reaction calorimeters. The Mettler-Toledo RC1 Reaction Calorimeter is one such device. The continuous rate of heat evolution (Qr), conversion, and time data can be collected for all emulsion polymerizations.

The molar enthalpy of polymerization typically has a distribution. Calorimetry can be used in conjunction with other online or offline approaches to track changes that take place during the polymerization process (Grotzinger et al., 2003). Industry commonly experiences polymerization rates of 10–4 mol/L H_2O/sec, albeit this varies depending on the formulation of the recipe and the capacity for heat removal.

4.6. PARTICLE SIZE AND SIZE DISTRIBUTION

Determining particle size and size distribution is a crucial aspect of the characterization of latex. The particle dimensions and distribution are tailored for individual latex formulations to fulfill their intended application demands (Chattopadhyay et al., 2001; Douglas et al., 2011). Binder particles within the range of 100 to 300 nm have been observed to interact with the surface of paper coatings, resulting in the production of desirable optical and mechanical properties. The dispersion rheology is influenced by particle size and size distribution, which impact its flow characteristics. The rheology of latex is a crucial factor in attaining a uniform surface in coatings-related contexts. In adhesive applications with high solids content exceeding 60%, it is necessary to employ bimodal or multimodal particle size distributions to ensure that the size remains manageable or monodisperse (Matyi et al., 1987).

This is despite the possibility of producing adhesives with a monodisperse particle size distribution, as previously reported in the literature.

A particle size distribution is typically observed in a singular sample of latex. The distribution's breadth is subject to variation across different latexes and is contingent upon the duration of the nucleation phase in proportion to the overall reaction. Secondary nucleation is responsible for the emergence of bimodal particle size distributions and multimodal. (Douglas et al., 1985; Verbeek, 2003).

The mean value is commonly used to express particle diameters. g mathematical formulas based on the distribution of particle sizes. These averages are widely used in various fields of science and engineering to describe the size of particles in a sample. The number-average diameter (D_n) represents the arithmetic mean of the particle diameters, while the volume-average diameter (D_v) and the weight-average diameter (D_w) take into account the volume and weight of each particle, respectively (Wallace, 2006). These measures provide valuable information about the physical properties and behavior of particles and can be used to optimize processes and design materials with specific characteristics (Figure 4.3):

$$D_n = \frac{\Sigma n_i d_i}{\Sigma n_i}$$

$$D_v = \left(\frac{\Sigma n_i d_i^3}{\Sigma n_i}\right)^{1/3}$$

$$D_w = \frac{\Sigma n_i d_i^4}{\Sigma n_i d_i^3}$$

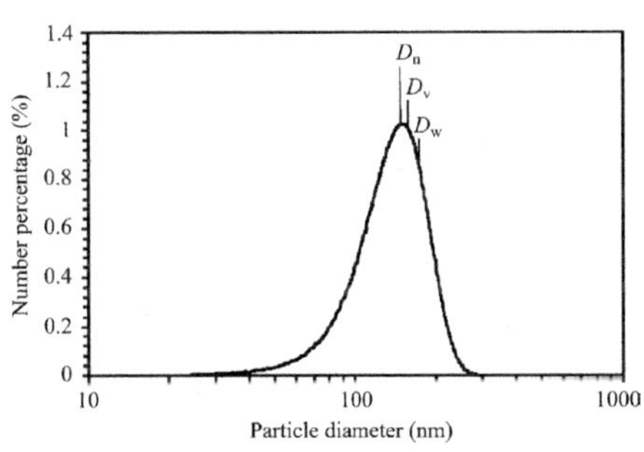

Figure 4.3. Typical particle size distribution for latex particles.

Source: Daniels, Creative Commons License.

where ni is the number of particles with actual diameter di. The number of particles per liter of water (Np) is then calculated from the volume-average diameter by:

$$N_p = \frac{6m}{\pi \rho D_v^3}$$

The variable "m" represents the ratio of polymer mass to water mass in the given context (González-Tello et al., 2008; Douglas et al., 1985).

Various methods are available for the measurement of particle sizes and size distributions. It is imperative to ensure that the samples are obtained with great care to accurately represent the entirety of the latex. Additionally, the samples must be meticulously prepared per the designated methodology. The various techniques for measuring particle size are classified into three primary categories. The initial approach is founded on microscopy, encompassing optical, electron, flow ultramicroscopy, and dark field microscopy techniques (Lescure et al., 1992). The subsequent category encompasses methodologies that employ light scattering, such as dynamic light scattering (DLS) and quasielastic light scattering (QELS). The third classification pertains to the motion of particles and includes various techniques such as fractional creaming, disc centrifuge sedimentation, microfiltration, hydrodynamic chromatography (HDC), and electro-zone sensing.

The optimal approach is contingent upon the scope of the size range under consideration, the level of exertion involved, and the degree of precision sought. Certain methodologies quantify size distributions, while others solely quantify mean diameters (Hoffman et al., 1992; Tanrisever et al., 1996). Particle size distributions smaller than 100 nm are commonly assessed through dynamic light scattering (DLS) or electron microscopy. However, it is worth noting that these methods can also be utilized for measuring larger size distributions. Particle diameters that exceed 1000 nm are typically determined through various methods such as sedimentation, optical microscopy, electro-zone sensing, or sieving.

4.7. PARTICLE MORPHOLOGY

The particle morphology pertains to the configuration or structure of the particles. Particles can exhibit various shapes such as uniform spheres, core-shell morphologies, hemispheres, domains or inclusions, non-spherical or irregular shapes, and inverted configurations where the core and shell pieces are reversed (Mikli et al., 2001). The morphology in question is governed by both thermodynamic and kinetic factors. Latex particles with porous structures have been synthesized in certain instances. Encapsulation of blowing agents, which are organic compounds characterized by low boiling temperatures, within latex particles has been observed in some instances. Upon undergoing thermal treatment, the blowing agent undergoes a phase transition into a gaseous state, causing the particle to increase in volume and adopt a hollow spherical morphology. Additional techniques for the preparation of hollow particles have been devised (Figure 4.4).

Did You Know?

The morphology of latex particles can vary greatly, ranging from spherical shapes to irregular and elongated structures, which can influence their performance in different applications.

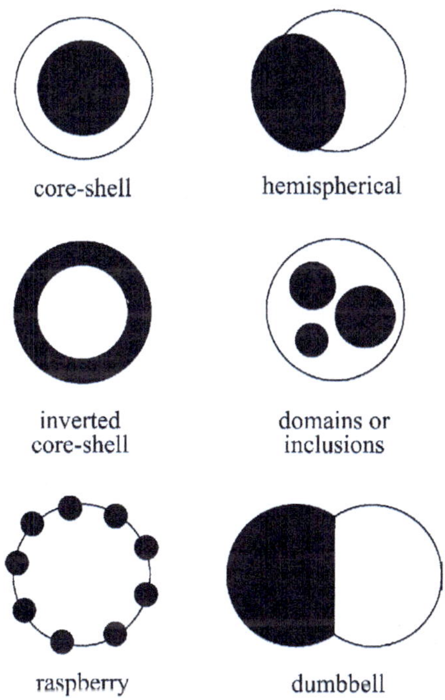

Figure 4.4. Potential morphologies of copolymer latex particles.

Source: Anderson, Creative Commons License.

The morphology of particles can significantly impact the performance of latex products. Latexes with multi-lobed particles have been utilized in coatings applications to increase viscosity due to their larger hydrodynamic volume compared to spherical particles with equivalent polymer mass.

The utilization of hollow particles in paper coatings is a common practice to enhance the optical characteristics and smoothness of the surface. Core-shell or domain-containing particles have been synthesized for the purpose of impact modification (DeCarlo et al., 2004; Sundberg & Durant, 2003). Furthermore, diverse microencapsulation methodologies have been utilized to encapsulate a broad range of substance intended for employment in the pharmaceutical, agricultural, and cosmetic sectors.

The examination of particle morphology is conducted through various techniques such as transmission electron microscopy (TEM), scanning electron microscopy (SEM), and atomic force microscopy (AFM). Optical microscopy (OM) can be employed in situations where the particle size attains a significant magnitude.

A drying process follows the application of latexes onto a substrate. Electron microscopy is performed in a state of high vacuum, utilizing an electron beam with a high accelerating voltage, which has the potential to cause degradation of the polymer particles (Li et al., 2019; Ortiz et al., 2020). In transmission electron microscopy (TEM), applying staining agents, such as osmium tetroxide, to the particles is frequently required to enhance contrast and facilitate their observation. Polymers with low glass transition temperatures exhibit a heightened vulnerability to softening, necessitating shadowing techniques to deduce the true dimensions and morphology of the particle.

In certain instances, it may be imperative to cryogenically preserve a pliable polymer specimen and subsequently section it into exceedingly thin slices in order to facilitate the visualization of internal architectures. The technique mentioned above is commonly referred to as cryo ultramicrotome. SEM is executed at reduced accelerating voltages; however, it is still susceptible to polymer degradation (Hamzehlou et al., 2019; Hawker, 1994). The SEM can generate particle images that exhibit a three-dimensional appearance, albeit with reduced magnification and resolution, when compared to the TEM. Figure 4.5 displays a transmission electron microscope image of a well-organized arrangement of uniform polystyrene particles generated through emulsion polymerization.

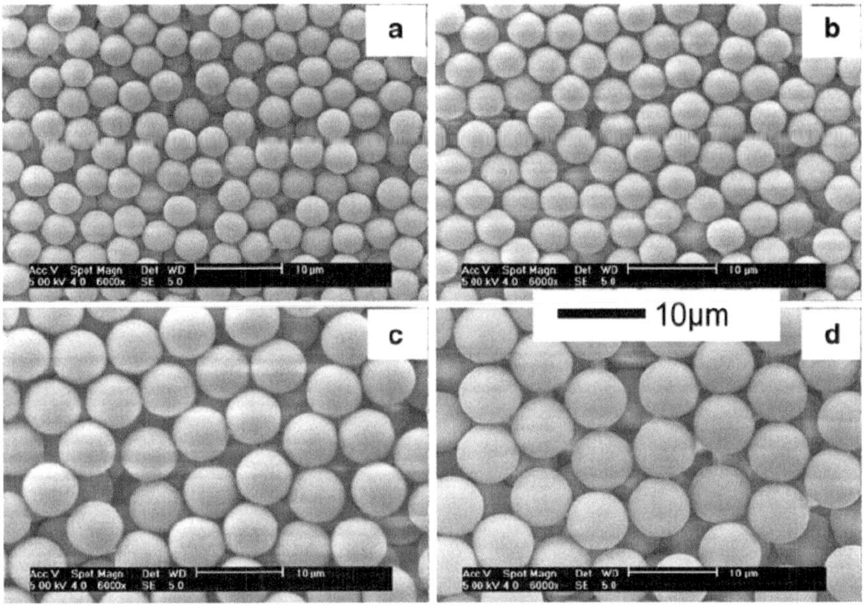

Figure 4.5. Monodisperse polystyrene latex particles made by emulsion polymerization are seen in a TEM micrograph.

Source: Anderson, Creative Commons License.

4.8. MOLECULAR WEIGHT

Determining the molecular weight, also called molar mass, is a crucial characteristic of a polymer. The durability of the polymer particle or film is closely associated with it. Maintaining a minimum molecular weight of 50,000 g/mol is recommended to ensure adequate chain entanglement and consequent toughness of the film (Honeyman et al., 1995). The interdiffusion rate for film formation in coatings applications is correlated with the length of the polymer chain. Pressure-sensitive adhesive (PSA) applications necessitate broad molecular weight distributions.

The molecular weight of a polymer depends on the radical flux that occurs during the polymerization process. Elevated concentrations of initiators result in an increased flux of radicals and a reduction in molecular weights. The reciprocal statement holds as well. Utilizing a chain transfer agent is a viable approach for regulating molecular weight (Nowalk et al., 2019; Fokou & Meier, 2008). Under semi-continuous feed conditions, a polymerization process can be conducted at a lower conversion rate to prevent the occurrence of gelation and branching. The determination of gel content in crosslinked systems can be achieved by means of swelling experiments.

The polymer sample exhibits a non-uniform molecular weight, wherein a distribution of polymer chain lengths is observed. Similar to the particle size distribution analysis, the molecular weight distribution is examined and quantified in relation to mean values (Chang & Liao, 1999). The measures that are most commonly utilized and considered significant are the number-average molecular weight (Mn), the weight-average molecular weight (Mw), and the viscosity-average molecular weight (Mv).

To determine the molecular weight of polymers present in latex samples, it is necessary to carry out a series of sequential steps. These steps include coagulation of the latex samples, cleaning the resultant coagulum, and ultimately dissolving the coagulum in a solvent appropriate for the specific method employed (Ito & Yamashita, 1965). The technique of Gel Permeation Chromatography (GPC) is commonly employed to determine the distribution of molecular weights. The Mark-Houwink equation is employed in viscometry as a basic technique for determining the viscosity-average molecular weight. Various techniques are available for obtaining precise molecular weight measurements, including ultracentrifugation, osmometry, scattering, and the relatively new matrix-assisted laser desorption ionization mass spectrometry (MALDI MS).

4.9. COPOLYMER COMPOSITION

When multiple monomers are utilized to prepare latexes, a copolymer is generated. The regulation of copolymer composition is achieved through the manipulation of comonomer quantities in the emulsion recipe and the comonomer reactivity ratios, which provide insight into the anticipated mode of interaction between distinct monomers, including random, alternating, or non-existent (resulting in the formation of two distinct homopolymers). The values of reactivity ratios, namely rA and rB, are obtained by blending the copolymer composition to the copolymerization equation through the utilization of Mayo-Lewis or Fineman-Ross plots. This fitting process is conducted across a spectrum of comonomer feed compositions (Chen, et al., 2009; Avgoustakis et al., 2003). Determining copolymer composition in emulsion polymerizations is typically achieved by inferring it from the measured comonomer feed composition using gas chromatography (GC) (220). The latexes can be subjected to drying, followed by dissolution of the polymer, and subsequently analyzed for copolymer composition through Fourier Transform Infrared (FTIR) or nuclear magnetic resonance (NMR) spectroscopy.

Despite having identical overall compositions, two distinct copolymer samples may exhibit a completely different sequence distribution, which refers to the arrangement of comonomer units within the polymer chain backbone. The formation of block copolymers, specifically in the context of styrene-butadiene (SBR) block copolymer rubbers, is a crucial consideration. NMR spectroscopy is frequently employed to ascertain the distribution of a sequence.

The copolymer composition of the latex plays a crucial role in determining the physical and chemical characteristics of the final product, particularly the glass transition temperature (Tg). The ability to control and customize the film formation ability, tackiness, and film strength of a copolymer can be achieved by altering its composition, specifically the glass transition temperature, thereby allowing for the attainment of specific requirements (Park, 1995; Franson & Peppas, 1983). The appropriate glass transition temperature range for PSA applications is low, specifically between -55 to -25° C. For latex paints, the range should be moderate, between 0 to 25℃, while for industrial coatings, the range should be relatively high, between 25 to 65°C.

The process of determining the glass transition temperature involves the utilization of differential scanning calorimetry (DSC) to measure the enthalpic changes of a polymer sample as it is subjected to heating (Qiao et al., 2005). This method has been documented in the literature. The point at which the heat capacity of the polymer undergoes a reduction is referred to as the glass transition temperature.

The copolymer composition also influences the latex product's hydrophobicity and flame retardancy. Incorporating suitable monomers through copolymerization can enhance water resistance and augment the flame retardancy of binders for carpet backing and nonwoven materials. Copolymers can be integrated into various proportions, for instance, 1% of acrylic acid in surface-functionalized polystyrene latexes or as high as 50% in acrylic copolymer latexes (Barluenga & Hernández-Olivares, 2004).

4.10. LATEX RHEOLOGY

The performance characteristics of latex are significantly influenced by its rheological properties, specifically its viscosity behavior, as evidenced by previous studies. The rheological properties of latex paints are of utmost significance as they play a crucial role in facilitating effortless stirring, spatter-free application, and uniform surface coverage (Senff et al., 1999; Maron et al., 1951). Latex-based paint is required to exhibit the ability to withstand sagging, particularly on vertical planes such as walls. The viscosity of the substance in question exhibits high values (104 cps) at low shear rates (1 s-1) and low values (100 cps) at high shear rates (103 s-1).

Remember

Understanding the rheological behavior of latex is crucial as it affects its processability, stability, and the properties of coatings or films produced from it.

Typically, the viscosity of latex is unsuitable for its intended application and necessitates augmentation by including a water-soluble viscosity modifier. The behavior of viscosity in latexes is significantly influenced by the interaction between particles. At elevated levels of solids concentration, the viscosity exhibits a rapid escalation (Panine et al., 2002). Conversely, particles of reduced size (at equivalent solids content) exhibit diminished inter-particle spacing, yielding heightened latex viscosity. The viscosity of latex is dependent on the particle size distribution. Latexes with high solids can be created as polydisperse or bimodal distribution latexes, where smaller particles occupy the spaces between larger particles. The impact of coagulum formation or the presence of water-soluble polymer on the rheology of latex, as well as the morphology of latex particles, can be significant (Allan, 1997; Alahapperuma & Glass, 1992).

Numerous commercial viscometers are accessible for the purpose of measuring the viscosity of latex. The cup variety is the most rudimentary, albeit the least precise. Capillary viscometry is employed for enhanced precision. Rotational viscometers, such as the Brookfield model, are commonly employed in order to quantify the shear stress across a broad spectrum of shear rates. Determining latex viscosity involves the computation of the quotient obtained by dividing the shear stress by the shear rate. The rheological behaviors, such as constant viscosity (Newtonian), shear-thickening or dilatant, and shear-thinning, can be identified by plotting shear stress or viscosity against shear rate. Hysteresis is the term used to describe the phenomenon whereby the viscosity of latex varies as a function of the rate of shear, increasing or decreasing (Felton, 2013). The phenomenon where the viscosity of latex decreases over time at a constant shear rate is referred to as thixotropy.

4.11. FILM FORMATION

Latexes are commonly utilized in various applications as coatings to protect substrates. The attainment of a cohesive film is the primary objective in various latex applications, including industrial coatings and latex paints. In instances of this nature, the process of film formation occurs over a period of several days. Waterborne emulsion polymers represent an environmentally favorable substitute for coatings that rely on solvents (Keddie, 1997). The coalescence of individual latex particles within the coating is a crucial process that leads to forming a continuous and unbroken polymer film. The film formation process advances through multiple stages. During the initial phase, the water content within the latex undergoes evaporation, leading to gradual contact between the individual latex particles. During the subsequent phase, capillary forces compress the latex particles, leading to the deformation of soft particles into hexagonal shapes. This process occurs while water evaporation persists. Polymer chain interdiffusion across particle interfaces may occur if the polymer chains are portable and the particles are soft.

Using coalescing agents is a viable approach to enhance the softening of latex polymers and expedite the film formation process, as indicated by a previous study. Elevating the temperature is known to expedite the process of film formation (Brinker et al., 1992; Thompson & Wood, 1981). Determining the minimum film-forming temperature (MFFT) involves the application of latex onto a substrate, followed by exposure of the latex to a temperature gradient and subsequent observation of the resulting film formation.

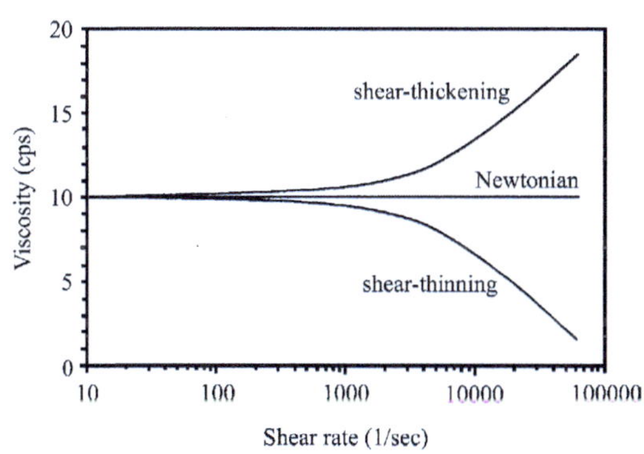

Figure 4.6. Common latex rheologies.

Source: C. D. Anderson and E. S. Daniels, Creative Commons License.

The MFFT is the temperature at which the film transitions from translucent to clear. The MFFT is contingent upon various factors, including the glass transition temperature, which denotes the temperature at which the rigid latex particles undergo a softening transition (Al-Saleh & Sundararaj, 2011). The polymer chain mobility and interdiffusion rate are also influenced by the molecular weight and crosslinking density of the latex particles (Figures 4.6 & 4.7).

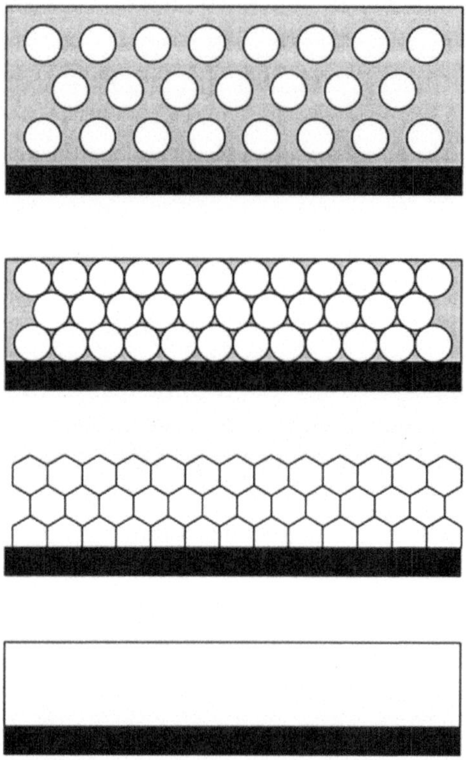

Figure 4.7. The film formation process.

Source: C. D. Anderson and E. S. Daniels, Creative Commons License.

4.12. MECHANICAL PROPERTIES

The calculation of the sample's toughness is based on the integration of the stress-strain curve area. The evaluation of impact and scratch resistance constitutes a crucial aspect in the assessment of coatings for various applications.

The dynamic mechanical analysis (DMA) technique is utilized to determine the viscoelastic characteristics of a given specimen. A sinusoidal deformation with a small amplitude is imparted to the latex film specimen, and the corresponding stress response is evaluated (Phung et al., 2003). The DMA enables the quantification of the storage modulus, which represents the energy retained in the sample, and the loss modulus, which corresponds to the energy dissipated from the sample as heat. The utilization of scanning probe microscopy has facilitated the investigation of mechanical interactions at the microscale.

4.13. PARTICLE SURFACE CHARACTERIZATION

Latexes exhibit a significant interfacial surface area at the particle interface, which accommodates diverse types of adsorbed molecules or chemically-bound species, such as surfactants for stabilization, charged end groups that arise from initiation, and various functional groups (e.g., carboxyl). The meticulous characterization of the surface layer is imperative regarding the durability and responsiveness of the latex particles, as indicated by previous studies. Numerous techniques for characterizing the surface of particles are frequently employed in academic research (Karim et al., 2011; Erdem et al., 2011). The latex particles undergo a serum replacement procedure as an initial step to eliminate dissolved components. Subsequently, ion exchange resins are employed to transform the counterions of chemically-bound ionic species into the H+ configuration. Conductometric titration is a method utilized to determine a particular parameter. The attributes of strength and resilience are highly sought after in various industrial applications involving latex. The durability of industrial coatings is crucial, as they are required to endure stress levels of approximately 1000 kPa without experiencing any form of failure. In certain instances, fabric binders are subjected to strains exceeding 50%. After forming a polymer film from latex, it is necessary to characterize its mechanical properties in an academic context (Yamauchi et al., 1989). The specimens are deposited onto a Teflon surface and subsequently extracted with caution upon desiccation. Subsequently, the specimen undergoes a tensile test wherein it is uniformly elongated until it fractures at a consistent velocity. The measurement of stress, which refers to the force exerted on a sample per unit of its cross-sectional area, is conducted in relation to the strain, which pertains to the degree of elongation. Young's modulus (E) can be determined by calculating the slope of the stress-strain curve. The ultimate strength of the specimen refers to the magnitude of stress at which it undergoes fracture through the integration of the quantity of charged entities within the latex. By taking into consideration the fact that the charges are evenly spread out across the entirety of the particle's surface area, it is possible to determine the surface charge density. Electrophoresis presents an alternative method for determining the surface charge density (Gómez-Morales et al., 2013; Khan et al., 2000). The surfactant surface coverage on latex particles is quantified through surfactant titration. The process of measuring surface tension in latex involves the addition of surfactant until the critical concentration of micelle (CMC) is attained, at which juncture the surface tension stabilizes. The surface coverage can be calculated based on the initial and critical micelle concentration levels of surfactant in the latex and the particle diameter. In certain industrial settings, achieving surface coverages greater than 50% in commercial latexes may be necessary to avoid coagulation (Yli-Urpo et al., 2005).

Activity 4.1

Imagine you are a chemist responsible for developing a new latex formulation for a specific application. Describe the step-by-step process you would follow to prepare and characterize the latex. Consider factors such as recipe formulation, selection of appropriate sensors for process control, characterization techniques for particle size and morphology analysis, and evaluation of film formation and mechanical properties.

4.14. SUMMARY

"Latex Preparation and Characterization" delves into the intricacies of preparing and characterizing latex materials. The chapter begins with an introduction, setting the stage for the following discussions. It then explores various aspects of latex preparation, including recipe formulation, feed strategies, sensors and process control, reactor agitation, and temperature control. These factors play crucial roles in determining the properties of the resulting latex. The chapter also focuses on the characterization of latex materials. It covers topics such as colloidal stability characterization, conversion tests, rate of polymerization, particle size and size distribution analysis, particle morphology examination, molecular weight determination, copolymer composition analysis, latex rheology, film formation evaluation, mechanical properties testing, and particle surface characterization.

REVIEW QUESTIONS

1. What are the key considerations in recipe formulation for latex preparation?
2. How can different feed strategies affect the properties of latex materials?
3. What is the role of sensors and process control in latex preparation?
4. How does reactor agitation impact the characteristics of latex particles?
5. Discuss the significance of temperature control in the latex preparation process.
6. What methods can be used to evaluate the colloidal stability of latex dispersions?
7. How is the rate of polymerization determined in latex systems?
8. Explain the techniques used for particle size and size distribution analysis in latex characterization.

REFERENCES

1. Alahapperuma, K., & Glass, J. E., (1992). The effect of latex particle size on the rheology of thickened dispersions. Progress in Organic Coatings, 21(1), 53–67.
2. Alex, R., & Nah, C., (2006). Preparation and characterization of organoclay - rubber nanocomposites via a new route with skim natural rubber latex. Journal of Applied Polymer Science, 102(4), 3277–3285.
3. Allan, M. L., (1997). Rheology of latex-modified grouts. Cement and Concrete Research, 27(12), 1875–1884.
4. Al-Saleh, M. H., & Sundararaj, U., (2011). Review of the mechanical properties of carbon nanofiber/polymer composites. Composites Part A: Applied Science and Manufacturing, 42(12), 2126–2142.
5. amai, H., Niino, K., & Suzawa, T., (1989). Surface characterization of styrene/sodium styrene sulfonate copolymer latex particles. Journal of Colloid and Interface Science, 131(1), 1–7.
6. Asua, J. M., (2004). Emulsion polymerization: From fundamental mechanisms to process developments. Journal of Polymer Science Part A: Polymer Chemistry, 42(5), 1025–1041.
7. Avgoustakis, K., Beletsi, A., Panagi, Z., Klepetsanis, P., Livaniou, E., Evangelatos, G., & Ithakissios, D. S., (2003). Effect of copolymer composition on the physicochemical characteristics, in vitro stability, and biodistribution of PLGA–mPEG nanoparticles. International Journal of Pharmaceutics, 259(1, 2), 115–127.
8. Bao, Z., Flanigan, C., Beyer, L., & Tao, J., (2015). Processing optimization of latex - compounded montmorillonite/styrene - butadiene rubber–polybutadiene rubber. Journal of Applied Polymer Science, 132(8).
9. Barluenga, G., & Hernández-Olivares, F., (2004). SBR latex modified mortar rheology and mechanical behavior. Cement and Concrete Research, 34(3), 527–535.
10. Bouschlicher, M. R., Rueggeberg, F. A., & Boyer, D. B., (2000). Effect of stepped light intensity on polymerization force and conversion in a photoactivated composite. Journal of Esthetic and Restorative Dentistry, 12(1), 23–32.
11. Boutti, S., Graillat, C., & McKenna, T. F., (2005). High solids content emulsion polymerization without intermediate seeds. Part II. In: Situ generation of bimodal latices. Polymer, 46(4), 1211–1222.
12. Brinker, C. J., Hurd, A. J., Schunk, P. R., Frye, G. C., & Ashley, C. S., (1992). Review of sol-gel thin film formation. Journal of Non-Crystalline Solids, 147, 424–436.

13. Burfield, D. R., & Ng, S. C., (1978). Persulphate initiated graft copolymerization of methacrylamide in natural rubber latex—I. The influence of non-rubber components. European Polymer Journal, 14(10), 789–792.
14. Calheiros, F. C., Daronch, M., Rueggeberg, F. A., & Braga, R. R., (2014). Effect of temperature on composite polymerization stress and degree of conversion. Dental Materials, 30(6), 613–618.
15. Calheiros, F. C., Daronch, M., Rueggeberg, F. A., & Braga, R. R., (2008). Influence of irradiant energy on degree of conversion, polymerization rate and shrinkage stress in an experimental resin composite system. Dental Materials, 24(9), 1164–1168.
16. Calheiros, F. C., Kawano, Y., Stansbury, J. W., & Braga, R. R., (2006). Influence of radiant exposure on contraction stress, degree of conversion and mechanical properties of resin composites. Dental Materials, 22(9), 799–803.
17. Ceska, G. W., (1974). Carboxyl - stabilized emulsion polymers. Journal of Applied Polymer Science, 18(8), 2493–2499.
18. Chainey, M., Wilkinson, M. C., & Hearn, J., (1982). Preparation of overcoated polymer latices by a "shot-growth" technique. Industrial & Engineering Chemistry Product Research and Development, 21(2), 171–176.
19. Chang, J. S., & Liao, P. H., (1999). Molecular weight control of a batch polymerization reactor: Experimental study. Industrial & Engineering Chemistry Research, 38(1), 144–153.
20. Chattopadhyay, D., Chakraborty, M., & Mandal, B. M., (2001). Dispersion polymerization of aniline using hydroxypropylcellulose as stabilizer: Role of rate of polymerization. Polymer International, 50(5), 538–544.
21. Chen, H., Choi, J. H., Salas-De La, C. D., Winey, K. I., & Elabd, Y. A., (2009). Polymerized ionic liquids: The effect of random copolymer composition on ion conduction. Macromolecules, 42(13), 4809–4816.
22. Chern, C. S., (2006). Emulsion polymerization mechanisms and kinetics. Progress in Polymer Science, 31(5), 443–486.
23. Covolan, V. L., Mei, L. H. I., & Rossi, C. L., (1997). Chemical modifications on polystyrene latex: Preparation and characterization for use in immunological applications. Polymers for Advanced Technologies, 8(1), 44–50.
24. Cui, X., Zhong, S., Gao, Y., & Wang, H., (2008). Preparation and characterization of emulsifier-free core–shell interpenetrating polymer network-fluorinated polyacrylate latex particles. Colloids and Surfaces A: Physicochemical and Engineering Aspects, 324(1–3), 14–21.

25. Daronch, M., Rueggeberg, F. A., De Goes, M. F., & Giudici, R., (2006). Polymerization kinetics of pre-heated composite. Journal of Dental Research, 85(1), 38–43.

26. DeCarlo, P. F., Slowik, J. G., Worsnop, D. R., Davidovits, P., & Jimenez, J. L., (2004). Particle morphology and density characterization by combined mobility and aerodynamic diameter measurements. Part 1: Theory. Aerosol Science and Technology, 38(12), 1185–1205.

27. Dimitratos, J., Elicabe, G., & Georgakis, C., (1994). Control of emulsion polymerization reactors. AIChE Journal, 40(12), 1993–2021.

28. Douglas, S. J., Illum, L., & Davis, S. S., (1985). Particle size and size distribution of poly (butyl 2-cyanoacrylate) nanoparticles. II. Influence of stabilizers. Journal of Colloid and Interface Science, 103(1), 154–163.

29. Douglas, S. J., Illum, L., Davis, S. S., & Krueter, J., (1984). Particle size and size distribution of poly (butyl-2-cyanoacrylate) nanoparticles: I. Influence of physicochemical factors. Journal of Colloid and Interface Science, 101(1), 149–158.

30. Dubois, E., Amram, B., Charmot, D., Menardo, C., & Ridoux, J. P., (1994). Latex emulsion polymerization and composition studies by Fourier transform Raman and infrared spectroscopies. In: 9th International Conference on Fourier Transform Spectroscopy (Vol. 2089, pp. 470, 471).

31. Erdem, B., Hunsicker, R. A., Simmons, G. W., Sudol, E. D., Dimonie, V. L., & El-Aasser, M. S., (2001). XPS and FTIR surface characterization of TiO_2 particles used in polymer encapsulation. Langmuir, 17(9), 2664–2669.

32. Erdoğan, S., Alpbaz, M., & Karagöz, A. R., (2002). The effect of operational conditions on the performance of batch polymerization reactor control. Chemical Engineering Journal, 86(3), 259–268.

33. Espeel, P., Goethals, F., Driessen, F., Nguyen, L. T. T., & Du Prez, F. E., (2013). One-pot, additive-free preparation of functionalized polyurethanes via amine–thiol–ene conjugation. Polymer Chemistry, 4(8), 2449–2456.

34. Felton, L. A., (2013). Mechanisms of polymeric film formation. International Journal of Pharmaceutics, 457(2), 423–427.

35. Flores-Cerrillo, J., & MacGregor, J. F., (2002). Control of particle size distributions in emulsion semibatch polymerization using mid-course correction policies. Industrial & Engineering Chemistry Research, 41(7), 1805–1814.

36. Fokou, P. A., & Meier, M. A., (2008). Acyclic triene metathesis polymerization with chain - stoppers: Molecular weight control in the synthesis of branched polymers. Macromolecular Rapid Communications, 29(19), 1620–1625.

37. Franson, N. M., & Peppas, N. A., (1983). Influence of copolymer composition on non - fickian water transport through glassy copolymers. Journal of Applied Polymer Science, 28(4), 1299–1310.
38. Gagnon, L., & MacGregor, J. F., (1991). State estimation for continuous emulsion polymerization. The Canadian Journal of Chemical Engineering, 69(3), 648–656.
39. Gómez-Morales, J., Iafisco, M., Delgado-López, J. M., Sarda, S., & Drouet, C., (2013). Progress on the preparation of nanocrystalline apatites and surface characterization: Overview of fundamental and applied aspects. Progress in Crystal Growth and Characterization of Materials, 59(1), 1–46.
40. Gonzaga, J. C. B., Meleiro, L. A. C., Kiang, C., & Maciel, F. R., (2009). ANN-based soft-sensor for real-time process monitoring and control of an industrial polymerization process. Computers & Chemical Engineering, 33(1), 43–49.
41. González-Tello, P., Camacho, F., Vicaria, J. M., & González, P. A., (2008). A modified Nukiyama–Tanasawa distribution function and a Rosin–Rammler model for the particle-size-distribution analysis. Powder Technology, 186(3), 278–281.
42. Grotzinger, C., Burget, D., Jacques, P., & Fouassier, J. P., (2003). Visible light induced photopolymerization: Speeding up the rate of polymerization by using co-initiators in dye/amine photoinitiating systems. Polymer, 44(13), 3671–3677.
43. Guo, Y., Zhang, J., Xie, P., Gao, X., & Luo, Y., (2014). Tailor-made compositional gradient copolymer by a many-shot RAFT emulsion polymerization method. Polymer Chemistry, 5(10), 3363–3371.
44. Guyot, A., Landfester, K., Schork, F. J., & Wang, C., (2007). Hybrid polymer latexes. Progress in Polymer Science, 32(12), 1439–1461.
45. Hamzehlou, S., Leiza, J. R., & Asua, J. M., (2016). A new approach for mathematical modeling of the dynamic development of particle morphology. Chemical Engineering Journal, 304, 655–666.
46. Hawker, C. J., (1994). Molecular weight control by a "living" free-radical polymerization process. Journal of the American Chemical Society, 116(24), 11185–11186.
47. Hawkett, B. S., Napper, D. H., & Gilbert, R. G., (1980). Seeded emulsion polymerization of styrene. Journal of the Chemical Society, Faraday Transactions 1: Physical Chemistry in Condensed Phases, 76, 1323–1343.
48. Hayashi, K., Ochi, H., & Okamura, S., (1964). Radiation - induced post polymerization of trioxane in the solid state. Journal of Polymer Science Part A: General Papers, 2(6), 2929–2946.

49. Hoffman, A. J., Yee, H., Mills, G., & Hoffmann, M. R., (1992). Photoinitiated polymerization of methyl methacrylate using Q-sized zinc oxide colloids. The Journal of Physical Chemistry, 96(13), 5540–5546.
50. Honeyman, C. H., Manners, I., Morrissey, C. T., & Allcock, H. R., (1995). Ambient temperature synthesis of poly (dichlorophosphazene) with molecular weight control. Journal of the American Chemical Society, 117(26), 7035–7036.
51. Issart, A., Godin, S., Preud'Homme, H., Bierla, K., Allal, A., & Szpunar, J., (2019). Direct screening of food packaging materials for post-polymerization residues, degradation products and additives by liquid extraction surface analysis nanoelectrospray mass spectrometry (LESA-nESI-MS). Analytica Chimica Acta, 1058, 117–126.
52. Ito, K., & Yamashita, Y., (1965). Copolymer composition and microstructure. Journal of Polymer Science Part A: General Papers, 3(6), 2165–2187.
53. Johnson, G. R., (1980). Effects of agitation during VCM suspension polymerization. Journal of Vinyl Technology, 2(3), 138–140.
54. Jönsson, J. E., Karlsson, O. J., Hassander, H., & Törnell, B., (2007). Semi-continuous emulsion polymerization of styrene in the presence of poly (methyl methacrylate) seed particles. Polymerization conditions giving core–shell particles. European Polymer Journal, 43(4), 1322–1332.
55. Karim, M. R., Rahman, M. A., Miah, M. A. J., Ahmad, H., Yanagisawa, M., & Ito, M., (2011). Synthesis of γ-alumina particles and surface characterization. The Open Colloid Science Journal, 4(1).
56. Keddie, J. L., (1997). Film formation of latex. Materials Science and Engineering: R: Reports, 21(3), 101–170.
57. Khan, M. A., Armes, S. P., Perruchot, C., Ouamara, H., Chehimi, M. M., Greaves, S. J., & Watts, J. F., (2000). Surface characterization of poly (3, 4-ethylenedioxythiophene)-coated latexes by X-ray photoelectron spectroscopy. Langmuir, 16(9), 4171–4179.
58. Lescure, F., Zimmer, C., Roy, D., & Couvreur, P., (1992). Optimization of polyalkylcyanoacrylate nanoparticle preparation: Influence of sulfur dioxide and pH on nanoparticle characteristics. Journal of Colloid and Interface Science, 154(1), 77–86.
59. Li, S., Li, J., Dong, M., Fan, S., Zhao, T., Wang, J., & Fan, W., (2019). Strategies to control zeolite particle morphology. Chemical Society Reviews, 48(3), 885–907.
60. Li, X., Liang, S., Wang, W. J., Li, B. G., Luo, Y., & Zhu, S., (2015). Model - based production of polymer chains having precisely designed end - to -

end gradient copolymer composition and chain topology distributions in controlled radical polymerization, A review. Macromolecular Reaction Engineering, 9(5), 409–417.

61. Liao, S. W., Cheng, Y. C., Gollakota, A. R., Chen, J. Y., & Shu, C. M., (2023). A thermal hazard risk evaluation of emulsion polymerization and vinyl acetate monomers in the latex manufacturing process. Journal of Loss Prevention in the Process Industries, 83, 105085.
62. Liu, Z., & Brooks, B. W., (1999). Inverse dispersion polymerization of acrylic acid initiated by a water-soluble redox pair: The role of drop mixing. Polymer, 40(9), 2181–2188.
63. Maron, S. H., Madow, B. P., & Krieger, I. M., (1951). Rheology of synthetic latex. II. Concentration dependence of flow in type V GR-S latex. Journal of Colloid Science, 6(6), 584–591.
64. Matyi, R. J., Schwartz, L. H., & Butt, J. B., (1987). Particle size, particle size distribution, and related measurements of supported metal catalysts. Catalysis Reviews Science and Engineering, 29(1), 41–99.
65. Mikli, V., Kaerdi, H., Kulu, P., & Besterci, M., (2001). Characterization of powder particle morphology. Proceedings of the Estonian Academy of Sciences: Engineering (Estonia), 7(1), 22–34.
66. Mülhaupt, R., (2003). Catalytic polymerization and post polymerization catalysis fifty years after the discovery of Ziegler's catalysts. Macromolecular Chemistry and Physics, 204(2), 289–327.
67. Noguchi, H., Yanase, N., Uchida, Y., & Suzuta, T., (1993). Preparation and characterization by thermal analysis of magnetic latex particles. Journal of Applied Polymer Science, 48(9), 1539–1547.
68. Nomura, M., Tobita, H., & Suzuki, K., (2005). Emulsion polymerization: Kinetic and mechanistic aspects. Polymer Particles, 1–128.
69. Nowalk, J. A., Fang, C., Short, A. L., Weiss, R. M., Swisher, J. H., Liu, P., & Meyer, T. Y., (2019). Sequence-controlled polymers through entropy-driven ring-opening metathesis polymerization: Theory, molecular weight control, and monomer design. Journal of the American Chemical Society, 141(14), 5741–5752.
70. Ortiz, D. G., Pochat-Bohatier, C., Cambedouzou, J., Bechelany, M., & Miele, P., (2020). Current trends in Pickering emulsions: Particle morphology and applications. Engineering, 6(4), 468–482.
71. Palin, W. M., Hadis, M. A., Leprince, J. G., Leloup, G., Boland, L., Fleming, G. J. P., & Watts, D. C., (2014). Reduced polymerization stress of MAPO-containing resin composites with increased curing speed,

degree of conversion and mechanical properties. Dental Materials, 30(5), 507–516.

72. Panine, P., Narayanan, T., Vermant, J., & Mewis, J., (2002). Structure and rheology during shear-induced crystallization of a latex suspension. Physical Review E, 66(2), 022401.
73. Park, T. G., (1995). Degradation of poly (lactic-co-glycolic acid) microspheres: Effect of copolymer composition. Biomaterials, 16(15), 1123–1130.
74. Pauer, W., (2018). Reactor concepts for continuous emulsion polymerization. Polymer Reaction Engineering of Dispersed Systems, I, 1–17.
75. Peula, J. M., Puig, J., Serra, J., De Las, N. F. J., & Hidalgo-Alvarez, R., (1994). Electrokinetic characterization and colloidal stability of polystyrene latex particles partially covered by IgG/a-CRP and m-BSA proteins. Colloids and Surfaces A: Physicochemical and Engineering Aspects, 92(1, 2), 127–136.
76. Phung, X., Groza, J., Stach, E. A., Williams, L. N., & Ritchey, S. B., (2003). Surface characterization of metal nanoparticles. Materials Science and Engineering: A, 359(1, 2), 261–268.
77. Piedra-Cascón, W., Krishnamurthy, V. R., Att, W., & Revilla-León, M., (2021). 3D printing parameters, supporting structures, slicing, and post-processing procedures of vat-polymerization additive manufacturing technologies: A narrative review. Journal of Dentistry, 109, 103630.
78. Plessis, C., Arzamendi, G., Leiza, J. R., Schoonbrood, H. A., Charmot, D., & Asua, J. M., (2001). Modeling of seeded semibatch emulsion polymerization of n-BA. Industrial & Engineering Chemistry Research, 40(18), 3883–3894.
79. Poehlein, G. W., & Dougherty, D. J., (1977). Continuous emulsion polymerization. Rubber Chemistry and Technology, 50(3), 601–638.
80. Qiao, M., Chen, D., Ma, X., & Liu, Y., (2005). Injectable biodegradable temperature-responsive PLGA–PEG–PLGA copolymers: Synthesis and effect of copolymer composition on the drug release from the copolymer-based hydrogels. International Journal of Pharmaceutics, 294(1, 2), 103–112.
81. Rawlings, J. B., & Ray, W. H., (1988). The modeling of batch and continuous emulsion polymerization reactors. Part I: Model formulation and sensitivity to parameters. Polymer Engineering & Science, 28(5), 237–256.

82. Sansatsadeekul, J., Sakdapipanich, J., & Rojruthai, P., (2011). Characterization of associated proteins and phospholipids in natural rubber latex. Journal of Bioscience and Bioengineering, 111(6), 628–634.
83. Schneider, L., F. J., Pfeifer, C. S., Consani, S., Prahl, S. A., & Ferracane, J. L., (2008). Influence of photoinitiator type on the rate of polymerization, degree of conversion, hardness and yellowing of dental resin composites. Dental Materials, 24(9), 1169–1177.
84. Schneider, M., Graillat, C., Guyot, A., & McKenna, T. F., (2002). High solids content emulsions. III. Synthesis of concentrated latices by classic emulsion polymerization. Journal of Applied Polymer Science, 84(10), 1916–1934.
85. Schork, F. J., & Ray, W. H., (1987). The dynamics of the continuous emulsion polymerization of methylmethacrylate. Journal of Applied Polymer Science, 34(3), 1259–1276.
86. Senff, H., Richtering, W., Norhausen, C., Weiss, A., & Ballauff, M., (1999). Rheology of a temperature sensitive core– shell latex. Langmuir, 15(1), 102–106.
87. Sosa, N., Peralta, R. D., López, R. G., Ramos, L. F., Katime, I., Cesteros, C., & Puig, J. E., (2001). A comparison of the characteristics of poly (vinyl acetate) latex with high solid content made by emulsion and semi-continuous microemulsion polymerization. Polymer, 42(16), 6923–6928.
88. Sundberg, D. C., & Durant, Y. G., (2003). Latex particle morphology, fundamental aspects: A review. Polymer Reaction Engineering, 11(3), 379–432.
89. Tanrisever, T., Okay, O., & Sönmezoğlu, I. Ç., (1996). Kinetics of emulsifier–free emulsion polymerization of methyl methacrylate. Journal of Applied Polymer Science, 61(3), 485–493.
90. Thompson, G. E., & Wood, G. C., (1981). Porous anodic film formation on aluminum. Nature, 290(5803), 230–232.
91. Twigt, F., Broekman, J., Piet, P., & German, A. L., (1993). Catalytic activity of cationic latices prepared according to the shot-growth method. European Polymer Journal, 29(5), 745–749.
92. Vallittu, P. K., Ruyter, I. E., & Buykuilmaz, S., (1998). Effect of polymerization temperature and time on the residual monomer content of denture base polymers. European Journal of Oral Sciences, 106(1), 588–593.
93. Vanderhoff, J. W., (1987). Semi-continuous emulsion polymerization. Future Directions in Polymer Colloids, 23–45.

94. Vanderhoff, J. W., (1993). Recent advances in the preparation of latexes. Chemical Engineering Science, 48(2), 203–217.

95. Verbeek, C. J. R., (2003). The influence of interfacial adhesion, particle size and size distribution on the predicted mechanical properties of particulate thermoplastic composites. Materials Letters, 57(13, 14), 1919–1924.

96. Wallace, L., (2006). Indoor sources of ultrafine and accumulation mode particles: Size distributions, size-resolved concentrations, and source strengths. Aerosol Science and Technology, 40(5), 348–360.

97. Wu, T. M., & Chen, E. C., (2008). Preparation and characterization of conductive carbon nanotube–polystyrene nanocomposites using latex technology. Composites Science and Technology, 68(10, 11), 2254–2259.

98. Xia, J., Gaynor, S. G., & Matyjaszewski, K., (1998). Controlled/ "living" radical polymerization. Atom transfer radical polymerization of acrylates at ambient temperature. Macromolecules, 31(17), 5958, 5959.

99. Xiong, M., Gu, G., You, B., & Wu, L., (2003). Preparation and characterization of poly (styrene butylacrylate) latex/nano - ZnO nanocomposites. Journal of Applied Polymer Science, 90(7), 1923–1931.

100. Xiong, M., Wu, L., Zhou, S., & You, B., (2002). Preparation and characterization of acrylic latex/nano - SiO_2 composites. Polymer International, 51(8), 693–698.

101. Xu, L., Xu, L., Dai, W. S., & Tsuboi, T., (2013). Preparation and characterization of a novel fluoro-silicone acrylate copolymer by semi-continuous emulsion polymerization. Journal of Fluorine Chemistry, 153, 68–73.

102. Yamauchi, H., Ishikawa, T., & Kondo, S., (1989). Surface characterization of ultramicro spherical particles of silica prepared by w/o microemulsion method. Colloids and Surfaces, 37, 71–80.

103. Yli-Urpo, H., Lassila, L. V., Närhi, T., & Vallittu, P. K., (2005). Compressive strength and surface characterization of glass ionomer cements modified by particles of bioactive glass. Dental Materials, 21(3), 201–209.

104. Zhang, Y., Zhuang, X., Gu, W., & Zhao, J., (2015). Synthesis of polyacrylonitrile nanoparticles at high monomer concentrations by AIBN-initiated semi-continuous emulsion polymerization method. European Polymer Journal, 67, 57–65.

CHAPTER

5 Environmentally Friendly Practices in Latex Production and Disposal

Contents

5.1. INTRODUCTION

Contemporary society exhibits a growing emphasis on sustainability and ecological accountability within diverse sectors. The latex industry is not an outlier. The utilization and elimination of latex, a multifaceted substance employed in various domains such as bonding, covering, and fabrics, may have noteworthy ecological ramifications (Chittella et al., 2021). Consequently, adopting eco-friendly measures in the production and disposal of latex is of paramount significance.

The significance of eco-friendly practices in latex production can be attributed to various pivotal factors. Initially, it aids in reducing the ecological ramifications linked to the acquisition and manipulation of primary resources (Qambrani et al., 2017). Latex can be obtained from natural sources, such as rubber trees, or can be synthesized using petrochemical-based monomers. Both procedures entail environmental factors such as land utilization, deforestation, consumption of energy, and the discharge of greenhouse gases. The adoption of sustainable sourcing practices for raw materials, such as the utilization of renewable resources or bio-based monomers, can effectively mitigate the carbon and ecological footprints associated with latex production. Furthermore, the implementation of eco-friendly techniques in the production of latex serves to promote the preservation and sustainable administration of natural resources (Mishra et al., 2019).

5.2. ENVIRONMENTAL IMPACT OF LATEX PRODUCTION

The production of latex has been found to have a notable ecological impact throughout its various lifecycle stages (Jawjit et al., 2015). The procurement of latex from rubber trees or the production of latex by means of petrochemical-derived monomers entails resource-intensive procedures and land utilization deliberations.

5.2.1. Overview of Environmental Impact

The process of latex production can potentially result in a noteworthy ecological footprint, spanning multiple phases ranging from the procurement of raw materials to the ultimate disposal of the finished goods. A comprehensive analysis of the environmental consequences linked to the production of latex facilitates the identification of crucial domains for enhancement and the implementation of ecologically sustainable methodologies (Skinner et al., 1997). The farming of rubber trees for the purpose of natural latex production is a significant environmental issue. The establishment of extensive rubber plantations has the potential to cause deforestation and the consequent loss of habitats, thereby causing disturbances to the local ecosystems as well as biodiversity. Moreover, the transformation of land for the purpose of rubber cultivation may lead to the depletion of carbon sinks which will give rise to the emission of greenhouse gases. The implementation of sustainable sourcing practices, which entail the promotion of accountable land use and reforestation initiatives, can serve as a means of reducing the negative effects associated with sourcing activities (Baykara, 2018). The environmental impact of latex production is also influenced by the energy consumption that occurs during the process. The conversion of unprocessed materials into usable forms, the formulation of emulsions, and the removal of moisture through drying procedures necessitate substantial energy consumption, frequently obtained from non-sustainable origins. The utilization of energy sources is a significant factor in the generation of carbon dioxide emissions as well as the exhaustion of non-renewable fossil fuel reserves. The implementation of energy-efficient technologies, incorporation of renewable energy sources, as well as optimization of production processes are effective measures in mitigating the carbon footprint associated with latex production (Menetrez, 2012) (Figure 5.1).

Figure 5.1. Harmful gases released by the synthetic rubber manufacturing industries.

Source: Premium flip-flops, Creative Commons License.

The utilization of water is a crucial environmental factor to be taken into account during the production of latex. The procurement of latex from rubber trees as well as the ensuing stages of processing necessitate significant quantities of water. In areas where there is a limited availability of water, the resultant pressure on the local water resources can lead to negative ecological consequences.

The adoption of water conservation techniques, such as the implementation of water-saving measures, recycling of water, and treatment of wastewater prior to discharge, are critical practices that serve to reduce the water footprint and promote responsible water management in the production of latex (Larsson, 2004).

The utilization of chemicals in the production of latex has been a subject of environmental apprehension. The utilization of emulsion stabilizers, cross-linking substances, and other chemicals during the process may comprise components that possess potential hazards to both human health and the environment. The implementation of appropriate chemical management practices and the replacement of perilous compounds with less harmful alternatives are imperative in mitigating the ecological ramifications of latex manufacturing.

The generation of waste during the production of latex presents significant environmental challenges. The proper disposal of sludge resulting from the treatment of wastewater, waste latex, and various other by-products is imperative to mitigate environmental contamination. The implementation of waste management methods that prefer recycling, reuse, and safe disposal can effectively mitigate the environmental impact associated with latex production (Drizo & Pegna, 2006).

5.2.2. Energy Consumption and Carbon Footprint

The environmental effect of latex production is significantly influenced by energy consumption. The processing of latex involves several stages such as extraction of raw materials, preparation of emulsion, drying, and curing, which necessitate significant energy consumption, frequently obtained from non-renewable sources (Le et al., 2010) (Figure 5.2).

The production of latex involves energy-intensive procedures that lead to the release of greenhouse gases, thereby contributing to the carbon footprint of the industry. The release of carbon dioxide (CO_2) into the atmosphere, predominantly resulting from the burning of non-renewable energy sources, is a significant issue of apprehension.

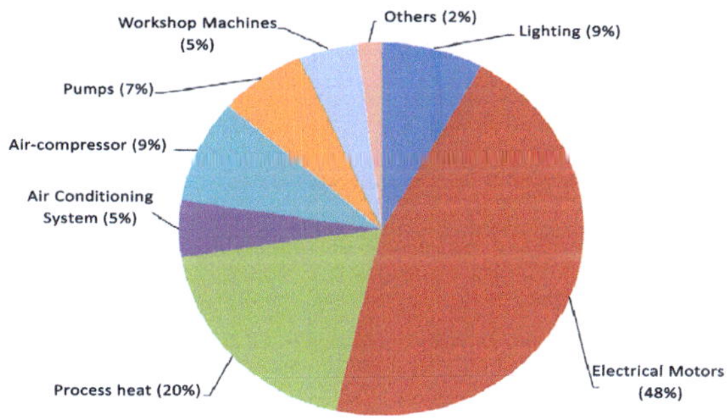

Figure 5.2. Percentage of energy consumption by end-use equipment of latex industry for year 2006.

Source: Saad Mekhilef, Creative Commons License.

Tip

Assess and optimize energy consumption during latex production to reduce the carbon footprint and minimize environmental impact.

The aforementioned emissions are a significant contributor to the phenomenon of climate change and the resultant global warming. In response to the environmental impact of latex production, the industry is progressively implementing energy-efficient methods and technologies to mitigate energy consumption as well as carbon footprint (Oluseyi et al., 2016).

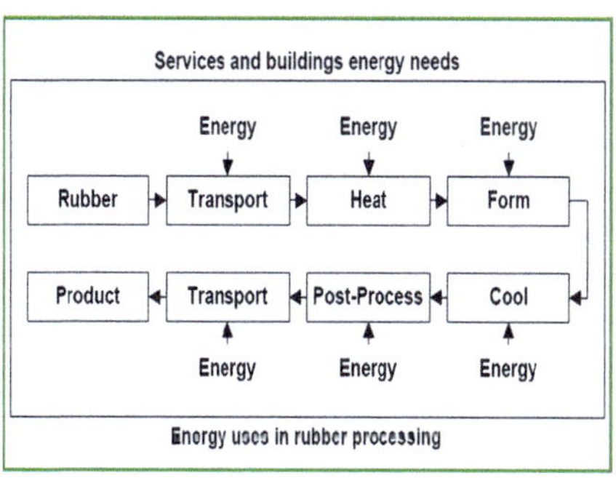

Figure 5.3. Steps in latex processing energy use.

Source: Saad Mekhilef, Creative Commons License.

Advanced drying techniques, including infrared and microwave drying, are utilized to minimize energy consumption in comparison to traditional drying methods. Furthermore, the optimization of production processes with the aim of minimizing energy loss, coupled with the implementation of energy management systems, can effectively mitigate energy consumption and its associated emissions (Figure 5.3).

In addition, the incorporation of sustainable energy sources, like wind or solar energy, into latex manufacturing plants can substantially mitigate the environmental impact (Fang et al., 2011). The employment of clean energy sources serves to alleviate the emission of greenhouse gases and fosters a more environmentally friendly energy portfolio. The latex industry can effectively curtail its energy consumption and carbon footprint by prioritizing energy efficiency, embracing renewable energy sources, and adopting sustainable energy management practices. The implementation of these steps not only facilitates the preservation of the environment but also augments the industry's enduring viability and adaptability in the wake of climate change adversities (Balogun et al., 2015).

5.2.3. Water Usage and Wastewater Management

The environmentally conscious methods employed in latex production necessitate careful attention to water usage as well as wastewater management. The extraction of latex, preparation of emulsion, and the following manufacturing stages entail substantial water consumption. The production of latex can potentially exert pressure on local water resources, especially in areas where shortages of water is a significant issue (Valta et al., 2014). The implementation of liable water management practices is of utmost importance in order to mitigate the environmental impact and guarantee the accessibility of water for alternative purposes.

The adoption of water-conserving technologies and practices is a viable strategy for mitigating water consumption (Burkhard et al., 2000). The strategies encompass enhancing water circulation mechanisms, repurposing and recycling water during the manufacturing process, and mitigating water wastage by means of leak identification and rectification. Through the implementation of these measures, the latex sector has the potential to substantially mitigate its water usage and mitigate stress on water reserves. The management of wastewater constitutes a crucial element in the pursuit of ecologically sustainable latex production. The production of latex entails a series of procedures that result in the generation of wastewater that is laden with diverse contaminants, including but not limited to latex particulates, chemicals, and additional impurities (Rad & Lewis, 2014). The appropriate management of wastewater is crucial in mitigating water body pollution and preserving ecological equilibrium.

The implementation of efficient wastewater treatment systems guarantees the elimination or remediation of pollutants prior to their release. Various treatment techniques, such as physical, chemical, as well as biological processes, can be employed to eliminate suspended particles, organic matter, as well as other contaminants (Tong & Elimelech, 2016). Effective as well as environmentally friendly wastewater treatment can be achieved through the implementation of advanced methods like filtration through membranes, activated carbon adsorption, as well as biological treatment. Apart from the treatment of wastewater, the reuse and recycling of water in the production procedure can additionally curtail water consumption and diminish the magnitude of wastewater produced. The implementation of closed-loop water systems, which involve the continuous treatment and recycling of water, has the potential to yield substantial water conservation benefits and mitigate environmental degradation (Capodaglio, 2017).

Furthermore, it is imperative to adhere to the pertinent regional guidelines and ecological criteria pertaining to the disposal of wastewater. The consistent surveillance of the quality of wastewater and adherence to established thresholds are imperative for upholding ecological soundness and safeguarding water reservoirs (Fillaudeau et al., 2016). Through the adoption of liable water usage practices, optimization of water management systems, and the introduction of successful wastewater treatment policies, the latex industry can effectively curtail its water footprint and mitigate the environmental consequences related to water usage as well as wastewater discharge. These practices have a dual benefit of promoting environmental sustainability and aiding in the preservation as well as preservation of water supplies for both current and future generations (Goh et al., 2019).

5.2.4. Chemical Usage and Waste Generation

The implementation of environmentally friendly procedures in latex production necessitates careful attention to chemical use and waste generation. The production of latex typically involves the utilization of diverse chemical agents for the stabilization of emulsions, cross-linking, and various other functional objectives. Improper storage and handling of chemicals may result in adverse impacts on both the environment and human health (Sari et al., 2022). The implementation of green chemistry methodologies can be a viable solution for the latex industry to mitigate the ecological consequences of chemical utilization. This entails the utilization of chemicals that are less hazardous and safer, minimizing the usage of dangerous substances, and optimizing procedures to decrease chemical consumption. Through the prioritization of ecologically sound chemicals, the industry may reduce the potential for environmental contamination and foster sustainable production methodologies (Ley, 2012) (Figure 5.4).

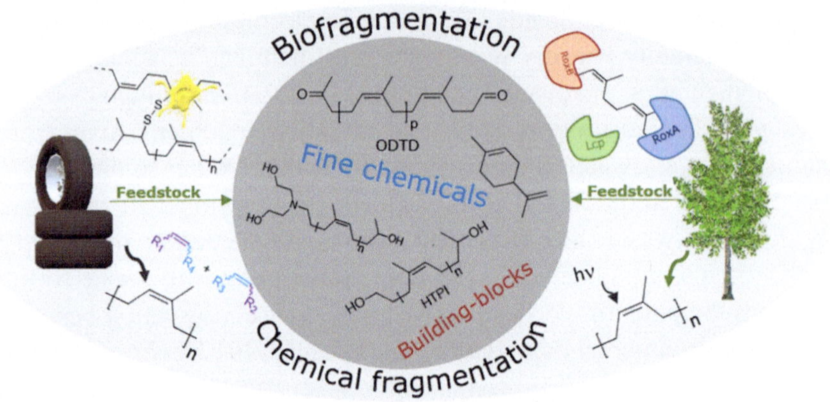

Figure 5.4. Natural rubber degradation products: Fine chemicals and reuse of rubber waste.

Source: Franciela Arenhart Soares, Creative Commons License.

It is imperative to undertake measures aimed at effectively regulating and appropriately disposing of chemical waste. The implementation of waste segregation practices, appropriate storage of chemicals in secure containers, and the establishment of protocols for properly handling and disposing of chemical waste are crucial measures to ensure safety and compliance in chemical management. Adhering to regulatory protocols and implementing optimal methodologies for the management of chemical waste guarantees the prevention of potential release of hazardous materials into the ecosystem (Narisetty et al., 2021). Apart from the utilization of chemicals, the generation of waste throughout the process of latex production is also a matter of environmental apprehension. The appropriate management of waste latex, sludge generated from the treatment of wastewater, and other by-products is crucial to mitigate potential challenges. The implementation of waste reduction strategies is crucial, whereby production processes are optimized to reduce waste generation and the use of raw materials is maximized. This approach has the potential to mitigate the cumulative environmental footprint and preserve finite resources.

In addition, there exist prospects for the latex industry to investigate avenues pertaining to the recycling and repurposing of waste materials (Onwukwe & Nwakaudu, 2012). The processing and utilization of waste latex may be observed in diverse applications, including the production of reclaimed rubber or its incorporation as a constituent in other commodities. The sludge generated from wastewater treatment has the potential to be utilized as a renewable energy source or transformed into valuable by-products following appropriate treatment procedures. The implementation of eco-friendly practices, optimization of chemical usage, and effective waste management strategies can significantly mitigate the environmental impact of chemical consumption and waste generation in the latex industry. These procedures not only facilitate a more hygienic and ecologically sound environment but also foster the preservation of resources and the promotion of long-term viability in the latex manufacturing process (Danalewich et al., 1998).

5.3. SUSTAINABLE LATEX PRODUCTION METHODS

The sustainable methods employed in latex production aim to reduce the ecological footprint of the process, while simultaneously upholding the standards of product performance and quality. The techniques employed encompass a range of production-related facets, such as the acquisition of raw materials, the adoption of environmentally-friendly chemical processes, the implementation of energy-efficient methodologies, and the curtailment of waste production (Onoji et al., 2016). The procurement of raw materials is a crucial element in ensuring sustainable latex production. The aforementioned pertains to the advocacy of conscientious land management, guaranteeing sustainable methods of cultivating rubber trees, and investigating alternative origins, like bio-based monomers that stem from renewable resources. Through the prioritization of environmentally friendly sourcing, the industry has the potential to mitigate deforestation, conserve biodiversity, and diminish the carbon footprint linked to the extraction of raw materials (Wen et al., 2017).

The implementation of green chemistry methodologies is a crucial component in achieving sustainable latex production (Jawjit et al., 2015). This entails the utilization of ecologically sustainable chemicals, curbing the employment of dangerous substances, and streamlining procedures to mitigate chemical consumption and the production of waste. Through the implementation of environmentally-friendly and sustainable substitutes, the industrial sector can effectively mitigate the adverse ecological effects of chemical inputs, all while ensuring optimal product efficacy. The implementation of energy-efficient production processes constitutes a crucial element in the pursuit of sustainable latex production. The process entails the employment of energy-efficient technologies, fine-tuning production parameters, and investigating sustainable energy sources to curtail the energy usage and environmental impact of latex manufacturing (Jamal et al., 2021). The implementation of energy-efficient practices within the industry has the potential to make a significant contribution towards the reduction of greenhouse gas emissions and the mitigation of the impacts of climate change.

In addition, the implementation of recycling and reducing waste programs is of utmost importance in ensuring sustainable latex production. Through the implementation of waste reduction strategies, optimization of production processes to enhance material utilization, and exploration of recycling and reprocessing alternatives for waste latex as well as by-products, the industry can effectively minimize waste and promote a circular economy approach (da Silva Menecucci et al., 2019). Environmentally friendly latex production methods have a primary objective of reducing the negative environmental effects throughout the entire production process, encompassing the sourcing of raw materials to the management of waste. The implementation of sustainable practices within the latex industry has the potential to decrease its ecological impact, preserve resources, and promote an environmentally friendly future.

5.3.1. Sourcing of Raw Materials

The sustainable procurement of essential components is a pivotal facet of eco-friendly latex manufacturing. The concept entails the implementation of methodologies that mitigate the ecological ramifications linked to the acquisition and manipulation of primary resources, all the while guaranteeing a dependable logistical network (Ghamisi et al., 2021). An effective strategy for achieving sustainable sourcing involves the promotion of responsible land use. The aforementioned activity pertains to the implementation of land stewardship techniques that prioritize the preservation of ecological systems and biodiversity. The strategy encompasses the avoidance of deforestation, safeguarding of vulnerable habitats, and conservation of natural resources.

The adoption of sustainable methods of land use not only aids in the reduction of ecological impact but also enhances the enduring feasibility of rubber tree plantations (Di Noi et al., 2020).

Apart from the land use aspect, the sustainable production of latex necessitates the assurance of sustainable cultivation practices for rubber trees. The aforementioned entails the advocacy of conscientious agricultural methodologies that prioritize the enhancement of soil quality, efficient water utilization, and preservation of biodiversity (Sherriff et al., 2022). The implementation of agroforestry methods, such as the incorporation of intercropping practices or the integration of plantations of rubber with natural habitats, can potentially bolster the resilience of ecosystems and promote the adoption of sustainable agricultural practices within the industry. The investigation of substitute origins of primary resources is an additional facet of sustainable procurement. The aforementioned pertains to the creation and application of bio-derived monomers sourced from sustainable materials. Bio-based monomers present a viable and eco-friendly substitute to conventional petrochemical-based monomers, thereby promoting sustainability. The industry can effectively decrease its carbon footprint by diminishing its dependence on non-renewable energy sources and advocating for the adoption of sustainable alternatives (Valentini et al., 2022).

Certifications and standards are crucial components in the context of sustainable sourcing. Several certification schemes, including the Forest Stewardship Council (FSC) as well as the Roundtable on Sustainable Palm Oil (RSPO), offer directives and prerequisites for sustainable sourcing methodologies (Bakshi et al., 2021). Compliance with these certifications guarantees responsible and sustainable sourcing of raw materials. The attainment of sustainable sourcing objectives is contingent upon the imperative of engaging together with community members and stakeholders. Fostering interaction with nearby communities, particularly those comprising small-scale agriculturalists, and upholding equitable and principled procedures across the entirety of the distribution network advances both societal and ecological sustainability (Kenzhaliyev et al., 2021).

Through the adoption of environmentally friendly sourcing procedures, the latex industry can effectively mitigate the environmental consequences that arise from the extraction of raw materials, conserve natural resources, and promote the welfare of indigenous communities (Gupta & Mahajan, 2015). The practice of sustainable sourcing is known to have a positive impact on both environmental stewardship and the long-term sustainability as well as endurance of the latex industry.

5.3.2. Green Chemistry Approaches

The implementation of green chemistry methodologies is a fundamental component in the achievement of sustainable latex manufacturing (Molnar et al., 2020). The primary objective of the organization is to mitigate the utilization of perilous substances, curtail the generation of waste, and optimize the processes to guarantee ecologically sound and sustainable practices all through the entire latex production cycle. The selection and utilization of more environmentally friendly and secure chemicals is a crucial element of green chemistry in the production of latex (Asiya et al., 2020). The process entails replacing perilous substances with eco-friendly alternatives that exhibit negligible effects on both human well-being and the ecosystem. Through the prioritization of less harmful and recyclable chemicals, the industry may successfully reduce the potential risks that are connected with chemical inputs (Figure 5.5).

Did You Know?

Green chemistry approaches in latex production focus on utilizing renewable resources, minimizing hazardous chemicals, and designing environmentally friendly processes, promoting sustainability.

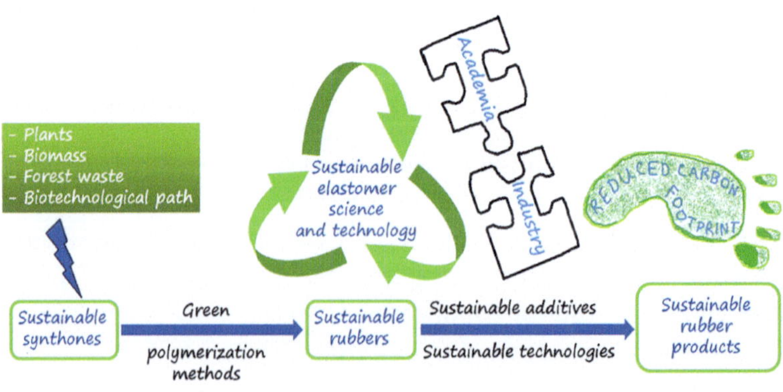

Figure 5.5. Concept of sustainable elastomer science and technology.

Source: Preetom Sarkar, Creative Commons License.

In addition, the implementation of production processes optimization aimed at reducing chemical consumption represents a crucial aspect of green chemistry (Shahid & Mohammad, 2013). The task at hand pertains to the formulation of emulsion stabilization techniques that are characterized by enhanced efficiency and reduced chemical consumption. The industry can mitigate potential environmental impacts and minimize waste generation by decreasing the total amount of chemicals utilized. Enhancing the environmental sustainability of latex formulations through their design and development is a pivotal facet of green chemistry (Bose et al., 2004). The development of low- volatile organic compounds (VOC) latex-based goods is aimed at reducing the emission of hazardous substances during their application and use. The utilization of low-VOC formulations is known to enhance the quality of indoor air and mitigate the potential adverse effects on human well-being and the environment.

Apart from the chemical aspects, green chemistry methodologies prioritize enhancing process efficiency to curtail energy usage and minimize waste production (Peng, 2002). The aforementioned encompasses the establishment and execution of energy-conserving manufacturing techniques, such as the enhancement of reaction parameters and the amelioration of thermal conduction procedures. The industry can decrease its carbon footprint as well as use of energy by minimizing its energy requirements. The adoption of waste reduction strategies is a crucial aspect of implementing green chemistry approaches (Darroudi et al., 2014). The aforementioned encompasses the optimization of processes with the aim of maximizing the utilization of materials while minimizing the generation of waste. Through the examination of recycling and reprocessing alternatives for waste latex and its by-products, the industry may efficiently decrease waste and encourage the implementation of a circular economy framework.

5.3.3. Energy-Efficient Production Processes

The attainment of environmentally friendly latex production necessitates the incorporation of energy efficiency as a fundamental component (Weinert et al., 2014). The implementation of energy-efficient production processes serves to mitigate energy consumption, diminish greenhouse gas emissions, and foster a more sustainable and ecologically conscious approach to the production of latex. An energy efficiency strategy involves an improvement of production parameters. The process entails the assessment and modification of parameters like temperature, pressure, reaction duration, and mixing circumstances to attain the intended level of product excellence while reducing energy consumption (Biel & Glock, 2016). Through the process of parameter fine-tuning, the industrial sector has the potential to mitigate energy waste and enhance the overall efficiency of their operations.

The implementation of cutting-edge technologies and equipment represents an additional approach towards achieving energy efficiency (Leobner et al., 2011). The implementation of advanced equipment, including energy-efficient reactors, exchangers of heat, and drying systems, can lead to a substantial decrease in energy usage during different phases of latex manufacturing. The adoption of systems for heat recovery has the potential to capture and recycle residual heat, leading to a reduction in energy consumption. The incorporation of sustainable energy sources constitutes a crucial element of energy-efficient manufacturing procedures. The integration of wind, solar, or biomass energy into the industrial sector can potentially diminish dependence on non-renewable sources of energy and mitigate its carbon emissions (Zhang et al., 2016). The implementation of on-site solar panels and wind turbines has the potential to produce sustainable and environmentally friendly energy to operate production facilities. Energy efficiency is also influenced by process optimization and automation. The deployment of process control systems and sophisticated monitoring technologies can facilitate the detection of energy-intensive zones and enable the optimization of operations in a targeted manner (Kovacic et al., 2013). Through diligent monitoring of energy usage and utilizing data-driven strategies, the industry can pinpoint opportunities for enhancement and successfully implement energy-conservation measures. The implementation of awareness and training initiatives aimed at employees is crucial in fostering energy-efficient practices. Incorporating the workforce into energy conservation initiatives, including the deactivation of equipment during periods of inactivity, the optimization of scheduling to minimize idle time, and the implementation of energy-saving measures, cultivates an energy-efficient culture within the organization (Kreitlein et al., 2015).

In addition, performing periodic evaluations and analyzes of energy consumption aids in the identification of potential avenues for energy conservation and facilitates the monitoring of advancements made towards achieving energy efficiency objectives (Joly et al., 2015). The aforementioned audits have the capability to detect regions where energy is being wasted, bring attention to possible enhancements or renovations, and offer valuable perspectives for continuous advancement. The implementation of energy-efficient production processes within the latex industry has the potential to decrease energy consumption, mitigate greenhouse gas emissions, and foster a more sustainable and ecologically responsible approach (Dietmair & Verl, 2009). The implementation of energy efficiency measures not only yields positive environmental outcomes, but also results in improved operational efficiency and decreased production costs over an extended period of time.

5.4. ECO-FRIENDLY LATEX ADDITIVES AND FORMULATIONS

The integration of eco-friendly latex ingredients and formulations is a crucial aspect in the attainment of environmentally friendly latex production. The additives and formulations are designed to mitigate the ecological consequences while preserving or augmenting the efficacy and characteristics of products based on latex (Hidayat et al., 2022).

5.4.1. Eco-Friendly Additives

The integration of eco-friendly additives is a pivotal factor in the attainment of sustainable latex production. These additives present viable alternatives to traditional additives, while simultaneously preserving or enhancing the performance and characteristics of latex-derived commodities, all in an environmentally responsible manner (Silva et al., 2020). The purpose of these additives is to mitigate the ecological consequences throughout the entire lifespan of the product, ranging from the acquisition of raw materials to the eventual disposal process (Figure 5.6).

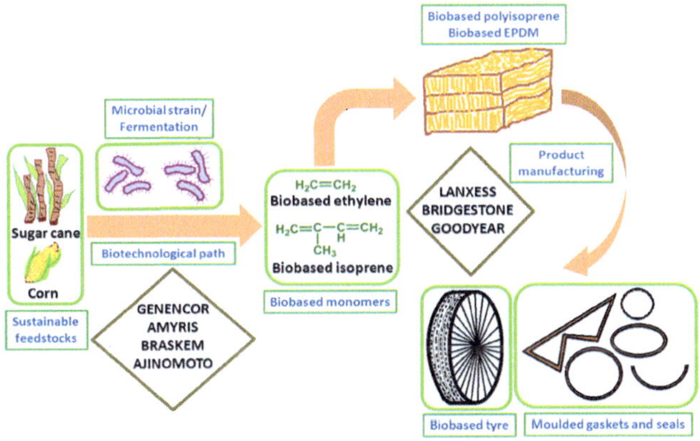

Figure 5.6. From sustainable feed-stocks to finished products.

Source: Preetom Sarkar, Creative Commons License.

One classification of environmentally friendly additives encompasses bio-based components that are obtained from sustainable sources (Gao et al., 2021). The utilization of bio-based plasticizers as well as surfactants presents a viable and eco-friendly substitute to the conventional petroleum-based additives that have been conventionally employed in the production of latex. The bio-based additives are obtained from various sources including plant oils, agricultural byproducts, and microbial fermentation. These alternatives offer comparable functional characteristics to their traditional equivalents, thereby decreasing the industry's dependence on non-renewable resources as well as mitigating carbon emissions. Eco-friendly latex additives necessitate the incorporation of natural thickening agents and stabilizers as a crucial component.

The aforementioned additives are sourced from sustainable materials, including extracts from plants or polysaccharides produced by microorganisms, and function as substitutes for synthetic agents that perform the functions of thickening and stabilizing. The incorporation of natural thickeners in latex formulations plays a crucial role in regulating viscosity and improving stability, thereby facilitating the attainment of the intended performance (Amanullah et al., 2016). Through the substitution of synthetic compounds with naturally-derived substitutes, the industrial sector has the potential to mitigate the ecological consequences that are commonly linked to the manufacturing, application, and elimination of synthetic additives.

Eco-friendly additives are comprised of essential constituents such as low- VOC as well as non-toxic formulations (Qaidi et al., 2021). VOCs are a class of chemicals that have the ability to evaporate at ambient temperatures and are known to be a significant contributor to atmospheric pollution, thereby posing potential health hazards. The industry can enhance indoor air quality and decrease emissions by creating latex products with reduced VOC content or by replacing VOC-containing chemicals with eco-friendly substitutes.

Incorporating non-toxic constituents in latex compositions reduces the likelihood of adverse effects on human well-being and the ecosystem, thereby promoting the development of safer and more environmentally-friendly commodities (Sarkar & Bhowmick, 2018). Thorough testing and evaluation are necessary for the development and execution of environmentally friendly additives. The performance criteria for the additives encompass their effectiveness, resilience, and compatibility with prevailing processes and applications. Furthermore, the sustainability features of the products, including their renewable sourcing, biodegradability, as well as overall environmental impact, are evaluated to ensure that they are consistent with sustainable goals (Praburanganathan et al., 2022).

Through the adoption of environmentally-friendly ingredients, the latex industry may successfully reduce its ecological impact, advance the cause of resource preservation, and cater to the escalating need for sustainable merchandise (Montazeri & Eckelman, 2018). The incorporation of these additives facilitates the industry's shift towards environmentally conscious practices, thereby promoting a sustainable and ecologically friendly future for the production and utilization of latex.

5.4.2. Bio-based Ingredients and Low-VOC Formulations

The integration of bio-based ingredients and low-VOC products is a crucial aspect of eco-friendly latex additives (Zareanshahraki & Mannari, 2018). These components provide sustainable substitutes for traditional additives, thereby mitigating the environmental consequences associated with latex production and utilization.

The utilization of bio-based ingredients obtained from renewable sources is crucial in advancing the sustainability of latex production (Robert et al., 2019). The constituents utilized in this context are procured from biomass, encompassing plant oils, agricultural residuals, or microbial fermentation. The incorporation of bio-based plasticizers, surfactants, as well as other additives can potentially decrease the industry's dependence on fossil fuel-derived constituents and facilitate the mitigation of greenhouse gas emissions. Bio-based additives offer comparable functionalities to their traditional counterparts, guaranteeing the preservation or enhancement of the desired performance attributes of latex products. The renewable nature of bio-based components is a notable advantage (Bland et al., 2008). Due to their origin from renewable sources, they present a more sustainable substitute for non-renewable fossil fuel resources. The incorporation of bio-based constituents also facilitates the advancement of a bioeconomy, which fosters the exploitation of biomass and endorses a more cyclical methodology to resource administration.

The incorporation of low-VOC formulations is a crucial component in the development of environmentally sustainable latex additives. VOCs are a class of chemical compounds that exhibit high vapor pressure at ambient temperature and can potentially pose a threat to both air quality and human health (Dai et al., 2015). The industry can enhance indoor air quality and decrease emissions by creating latex products with reduced VOC content. VOC formulations are attained through the substitution or reduction of the utilization of volatile solvents along with other chemicals that emit VOCs with eco-friendly substitutes (Dai et al., 2015). The available options comprise of aqueous-based systems or non-hazardous solvents that exhibit negligible effects on the atmosphere and human well-being.

The utilization of low-VOC formulations holds paramount importance in scenarios where emissions as well as indoor air quality pose significant apprehensions, such as in the case of adhesives, coatings, as well as sealants. The adoption of low-VOC alternatives by manufacturers can enable them to comply with regulatory standards, improve the safety of their workplace, and promote a more salubrious environment for their consumers (Adamová et al., 2019). The process of creating and executing bio-based ingredients as well as low-VOC formulations necessitates meticulous evaluation of their performance, reliability, and environmental ramifications. Thorough testing and assessment procedures are implemented to ensure that these environmentally conscious additives satisfy the essential performance criteria while simultaneously aligning with sustainability objectives.

The integration of bio-based constituents and low-VOC formulations in the latex sector can potentially diminish its ecological impact, reduce reliance on non-renewable resources, and offer consumers more sustainable and eco-friendly latex commodities (Gustafsson & Börjesson, 2007). The incorporation of these additives serves as a testament to the industry's dedication towards sustainable methodologies and fosters an environmentally conscious and salubrious prospect.

5.4.3. Natural Thickeners and Stabilizers

The utilization of natural thickeners and stabilizers is a crucial aspect of eco-friendly latex additives, which provide sustainable substitutes for synthetic alternatives. These natural components are capable of maintaining or even improving the efficiency and durability of latex-based products (Nakamatsu et al., 2017). Natural thickeners and stabilizers, sourced from renewable materials like extracts from plants or microbial polysaccharides, are utilized to regulate viscosity and improve the long-term stability of latex formulations. The incorporation of additives into latex-based products serves to enhance their flow properties, mitigate sedimentation, and guarantee their intended performance and functionality.

Remember

Consider using natural thickeners and stabilizers in latex formulations as eco-friendly alternatives, reducing the reliance on synthetic additives and improving the sustainability profile of the final product.

The biodegradability of natural thickeners as well as stabilizers is considered to be an advantageous feature (Promlok et al., 2009). In contrast to their synthetic counterparts, natural additives exhibit a greater propensity for biodegradation, thereby mitigating the potential ecological ramifications that may arise from their utilization and disposal.

In addition, natural substances that have thickening and stabilizing properties are frequently regarded as safer options. Commonly, these substances are considered harmless and possess a reduced likelihood of causing detrimental impacts on both human health and the ecosystem (Selvi et al., 2019). The utilization of natural ingredients can serve as a means for the industry to mitigate potential hazards and foster safer working conditions. Apart from their utilitarian advantages, naturally occurring thickeners and stabilizers play a significant role in enhancing the ecological viability of latex manufacturing (Krickl et al., 2017). The employment of sustainable resources diminishes reliance on fossil fuel-derived supplements and advances the preservation of finite resources. Thorough testing and evaluation are necessary for the development and execution of natural thickeners as well as stabilizers to guarantee their compatibility with diverse latex formulations and uses. The preservation or enhancement of the intended performance attributes of latex-based goods, including but not limited to viscosity, stability, as well as shelf-life, is guaranteed while simultaneously adhering to sustainability objectives. The integration of natural thickeners as well as stabilizers within the latex industry has the potential to mitigate its ecological footprint, foster the preservation of resources, and offer consumers latex products that are more sustainable and ecologically sound. The incorporation of these additives serves as a testament to the industry's dedication towards sustainable practices and facilitates the progression towards a more environmentally conscious and sustainable future.

5.4.4. Impact on Latex Performance and Properties

The incorporation of environmentally friendly additives, like bio-derived components, formulations with low levels of VOCs, and naturally sourced thickeners and stabilizers, into the manufacturing process of latex has a discernible influence on the characteristics and functionality of products that are based on latex. The integration of bio-based constituents into latex compositions has the potential to preserve or augment the operational attributes of the ultimate commodity. The aforementioned additives frequently demonstrate comparable functionalities to their traditional counterparts, thereby guaranteeing the preservation of the desired characteristics, including but not limited to adhesion, flexibility, and durability. Bio-based additives have the potential to enhance specific performance attributes, including heightened compatibility with different substances and augmented opposition to environmental factors (Poovaneshvaran et al., 2020).

The utilization of low-VOC formulations guarantees that latex products conform to rigorous environmental regulations and standards while maintaining their performance. Formulations with low VOCs demonstrate similar performance to those with high VOCs, thereby ensuring exceptional adhesion, strength, and various other favorable characteristics, while simultaneously reducing potential health hazards and environmental consequences. The incorporation of natural thickeners as well as stabilizers into latex formulations enhances their processing and performance by contributing to stability and viscosity control. The incorporation of additives in latex products serves to prevent sedimentation, promote uniformity, and augment the general stability of the products, thereby guaranteeing a consistent level of quality and performance.

The evaluation of the effects of environmentally sustainable additives on the performance and characteristics of latex is a critical aspect to be taken into account during the formulation and adoption of such additives. Thorough testing and evaluation procedures are implemented to guarantee the attainment of the intended functional attributes, like mechanical properties, film development, and compatibility. The aforementioned statement highlights the significance of eco-friendly additives in meeting or surpassing the performance benchmarks of traditional additives, with the added advantage of promoting environmental sustainability. The incorporation of environmentally friendly additives presents an opportunity for the latex industry to attain equilibrium between optimal performance and sustainable practices. The incorporation of additives in latex-based products serves the dual purpose of preserving or augmenting their intended characteristics while also aligning with the industry's objective of mitigating its ecological footprint and fostering greater sustainability measures (Ng et al., 2022).

Activity 5.1

Imagine you are a production manager in a latex manufacturing plant. Describe three specific steps you would take to minimize the environmental impact of latex production and promote environmentally friendly practices in your facility.

5.5. SUMMARY

The chapter titled "Environmentally Friendly Practices in Latex Production and Disposal" delves into diverse facets of advancing sustainability and mitigating the ecological ramifications of latex production and disposal. The text commences by emphasizing the significance of ecologically sustainable practices within the latex industry and presents a synopsis of the chapter's contents. The present chapter explores the environmental ramifications of latex production, encompassing the utilization of energy, water, as well as chemicals, and examines the difficulties related to producing waste and disposal. The statement underscores the significance of procuring raw materials in a sustainable manner, which entails adopting responsible land use as well as cultivation practices, alongside the creation of bio-based constituents and formulations with low VOCs. Additionally, the chapter investigates the involvement of green chemistry methodologies in the production of latex, with particular emphasis on the incorporation of environmentally sustainable additives and compositions. The text underscores the advantages of incorporating bio-based constituents, natural thickeners, as well as stabilizers, alongside low-VOC formulations, to sustain or enhance the efficacy of latex, while simultaneously mitigating ecological damage.

REVIEW QUESTIONS

1. What are some of the environmental impacts associated with latex production?
2. How does latex production contribute to energy consumption and carbon footprint?
3. Discuss the importance of water usage and wastewater management in latex production.
4. What are the concerns related to chemical usage and waste generation in latex production?

5. What are sustainable methods for latex production, and how do they contribute to environmental conservation?
6. Explain the concept of green chemistry approaches in latex production.
7. How can energy-efficient production processes be implemented in latex manufacturing?
8. What are eco-friendly additives and their significance in latex formulations?
9. Discuss the use of bio-based ingredients and low-VOC formulations in environmentally friendly latex production.
10. How do natural thickeners and stabilizers contribute to sustainable latex formulations.

REFERENCES

1. Adamová, T., Hradecký, J., & Prajer, M., (2019). VOC emissions from spruce strands and hemp shive: In search for a low emission raw material for bio-based construction materials. Materials, 12(12), 2026.
2. Amanullah, M., Ramasamy, J., Al-Arfaj, M. K., & Aramco, S., (2016). Application of an indigenous eco-friendly raw material as fluid loss additive. Journal of Petroleum Science and Engineering, 139, 191–197.
3. Asiya, S. I., Pal, K., Kralj, S., El-Sayyad, G. S., De Souza, F. G., & Narayanan, T., (2020). Sustainable preparation of gold nanoparticles via green chemistry approach for biogenic applications. Materials Today Chemistry, 17, 100327.
4. Bakshi, P., Pappu, A., & Gupta, M. K., (2021). A review on calcium-rich industrial wastes: A sustainable source of raw materials in India for civil infrastructure—Opportunities and challenges to bond circular economy. Journal of Material Cycles and Waste Management, 1–14.
5. Balogun, V. A., Kirkwood, N., & Mativenga, P. T., (2015). Energy consumption and carbon footprint analysis of fused deposition modeling: A case study of RP Stratasys dimension SST FDM. Energy Consumption and Carbon Footprint Analysis of Fused Deposition Modeling: A Case Study of RP Stratasys Dimension SST FDM, 6(8), 1–6.
6. Baykara, S. Z., (2018). Hydrogen: A brief overview on its sources, production, and environmental impact. International Journal of Hydrogen Energy, 43(23), 10605–10614.
7. Biel, K., & Glock, C. H., (2016). Systematic literature review of decision support models for energy-efficient production planning. Computers & Industrial Engineering, 101, 243–259.
8. Bland, K. E., Gromala, D. S., & Brooks, R. T., (2008). Sustainability attributes of bio-based structural building products. In: Structures Congress 2008: Crossing Borders (Vol. 1, pp. 1–8)
9. Bose, A. K., Pednekar, S., Ganguly, S. N., Chakraborty, G., & Manhas, M. S., (2004). A simplified green chemistry approach to the biginelli reaction using 'grindstone chemistry.' Tetrahedron Letters, 45(45), 8351–8353.
10. Burkhard, R., Deletic, A., & Craig, A., (2000). Techniques for water and wastewater management: A review of techniques and their integration in planning. Urban Water, 2(3), 197–221.
11. Capodaglio, A. G., (2017). Integrated, decentralized wastewater management for resource recovery in rural and peri-urban areas. Resources, 6(2), 22.

12. Chittella, H., Yoon, L. W., Ramarad, S., & Lai, Z. W., (2021). Rubber waste management: A review on methods, mechanism, and prospects. Polymer Degradation and Stability, 194, 109761.
13. Da Silva, M. C., Mucellini, K. L., De Oliveira, M. M., Higashi, B., De Almeida, R. T. R., Porto, C., & De Oliveira, A. J. B., (2019). Latex from Tabernaemontana catharinensis (A. DC)—Apocynaceae: An alternative for the sustainable production of biologically active compounds. Industrial Crops and Products, 129, 74–84.
14. Dai, J., Ma, S., Wu, Y., Zhu, J., & Liu, X., (2015). High bio-based content waterborne UV-curable coatings with excellent adhesion and flexibility. Progress in Organic Coatings, 87, 197–203.
15. Danalewich, J. R., Papagiannis, T. G., Belyea, R. L., Tumbleson, M. E., & Raskin, L., (1998). Characterization of dairy waste streams, current treatment practices, and potential for biological nutrient removal. Water Research, 32(12), 3555–3568.
16. Darroudi, M., Sabouri, Z., Oskuee, R. K., Zak, A. K., Kargar, H., & Abd, H. M. H. N., (2014). Green chemistry approach for the synthesis of ZnO nanopowders and their cytotoxic effects. Ceramics International, 40(3), 4827–4831.
17. Di Noi, C., Ciroth, A., Mancini, L., Eynard, U., Pennington, D., & Blengini, G. A., (2020). Can S-LCA methodology support responsible sourcing of raw materials in EU policy context? The International Journal of Life Cycle Assessment, 25, 332–349.
18. Dietmair, A., & Verl, A., (2009). A generic energy consumption model for decision making and energy efficiency optimization in manufacturing. International Journal of Sustainable Engineering, 2(2), 123–133.
19. Drizo, A., & Pegna, J., (2006). Environmental impacts of rapid prototyping: An overview of research to date. Rapid Prototyping Journal, 12(2), 64–71.
20. Fang, K., Uhan, N., Zhao, F., & Sutherland, J. W., (2011). A new approach to scheduling in manufacturing for power consumption and carbon footprint reduction. Journal of Manufacturing Systems, 30(4), 234–240.
21. Fillaudeau, L., Blanpain-Avet, P., & Daufin, G., (2006). Water, wastewater, and waste management in brewing industries. Journal of Cleaner Production, 14(5), 463–471.
22. Gao, C., Wolff, S., & Wang, S., (2021). Eco-friendly additive manufacturing of metals: Energy efficiency and life cycle analysis. Journal of Manufacturing Systems, 60, 459–472.
23. Ghamisi, P., Shahi, K. R., Duan, P., Rasti, B., Lorenz, S., Booysen, R., & Gloaguen, R., (2021). The potential of machine learning for a more

responsible sourcing of critical raw materials. IEEE Journal of Selected Topics in Applied Earth Observations and Remote Sensing, 14, 8971–8988.

24. Goh, H. Y., Wong, W. W. C., & Ong, Y. Y., (2019). A study to reduce chemical waste generated in chemistry teaching laboratories. Journal of Chemical Education, 97(1), 87–96.
25. Gupta, P., & Mahajan, A., (2015). Green chemistry approaches as sustainable alternatives to conventional strategies in the pharmaceutical industry. RSC Advances, 5(34), 26686–26705.
26. Gustafsson, L. M., & Börjesson, P., (2007). Life cycle assessment in green chemistry: A comparison of various industrial wood surface coatings. The International Journal of Life Cycle Assessment, 12, 151–159.
27. Hidayat, W., Aprilliana, N., Asmara, S., Bakri, S., Hidayati, S., Banuwa, I. S., & Iswanto, A. H., (2022). Performance of eco-friendly particleboard from agro-industrial residues bonded with formaldehyde-free natural rubber latex adhesive for interior applications. Polymer Composites, 43(4), 2222–2233.
28. Jamal, H., Lyne, A., Ashley, P., & Duane, B., (2021). Non-sterile examination gloves and sterile surgical gloves: Which are more sustainable? Journal of Hospital Infection, 118, 87–95.
29. Jawjit, W., Pavasant, P., & Kroeze, C., (2015). Evaluating environmental performance of concentrated latex production in Thailand. Journal of Cleaner Production, 98, 84–91.
30. Joly, M., Antonetti, Y., Python, M., Lazo, M. G., Gascou, T., Hessler-Wyser, A., & Schüler, A., (2014). Selective solar absorber coatings on receiver tubes for CSP–energy-efficient production process by sol-gel dip-coating and subsequent induction heating. Energy Procedia, 57, 487–496.
31. Kenzhaliyev, B., Imangalieva, L., Manapova, A., & Azlan, M., (2021). Kaolinite clays as a source of raw materials for the aluminum industry of the Republic of Kazakhstan. Kompleksnoe Ispolzovanie Mineralnogo Syra= Complex use of Mineral Resources, 319(4), 5–12.
32. Kovacic, I., Orehounig, K., Mahdavi, A., Bleicher, F., Dimitrou, A. A., & Waltenbereger, L., (2013). Energy efficient production–interdisciplinary, systemic approach through integrated simulation. Strojarstvo: Časopis za Teoriju i Praksu u Strojarstvu, 55(1), 17–34.
33. Kreitlein, S., Höft, A., Schwender, S., & Franke, J., (2015). Green factories Bavaria: A network of distributed learning factories for energy efficient production. Procedia CIRP, 32, 58–63.

34. Krickl, S., Touraud, D., & Kunz, W., (2017). Investigation of ethanolamine stabilized natural rubber latex from Taraxacum Kok-Saghyz and from Hevea Brasiliensis using zeta-potential and dynamic light scattering measurements. Industrial Crops and Products, 103, 169–174.
35. Larsson, C. M., (2004). The FASSET Framework for assessment of environmental impact of ionizing radiation in European ecosystems—An overview. Journal of Radiological Protection, 24(4A), A1.
36. Le, K., Bilgir, O., Bianchini, R., Martonosi, M., & Nguyen, T. D., (2010). Managing the cost, energy consumption, and carbon footprint of internet services. ACM SIGMETRICS Performance Evaluation Review, 38(1), 357–358.
37. Leobner, I., Ponweiser, K., Neugschwandtner, G., & Kastner, W., (2011). Energy efficient production-a holistic modeling approach. In: 2011 World Congress on Sustainable Technologies (WCST) (pp. 62–67).
38. Ley, S. V., (2012). On being green: Can flow chemistry help? The Chemical Record, 12(4), 378–390.
39. Menetrez, M. Y., (2012). An overview of algae biofuel production and potential environmental impact. Environmental Science & Technology, 46(13), 7073–7085.
40. Mishra, S., Chowdhary, P., & Bharagava, R. N., (2019). Conventional methods for the removal of industrial pollutants, their merits, and demerits. Emerging and Eco-friendly Approaches for Waste Management, 1–31.
41. Molnar, M., Lončarić, M., & Kovač, M., (2020). Green chemistry approaches to the synthesis of coumarin derivatives. Current Organic Chemistry, 24(1), 4–43.
42. Montazeri, M., & Eckelman, M. J., (2018). Life cycle assessment of UV-Curable bio-based wood flooring coatings. Journal of Cleaner Production, 192, 932–939.
43. Nakamatsu, J., Kim, S., Ayarza, J., Ramírez, E., Elgegren, M., & Aguilar, R., (2017). Eco-friendly modification of earthen construction with carrageenan: Water durability and mechanical assessment. Construction and Building Materials, 139, 193–202.
44. Narisetty, V., Cox, R., Willoughby, N., Aktas, E., Tiwari, B., Matharu, A. S., & Kumar, V., (2021). Recycling bread waste into chemical building blocks using a circular biorefining approach. Sustainable Energy & Fuels, 5(19), 4842–4849.
45. Ng, J. W., Othman, N., & Yusof, N. H., (2022). Various coagulation techniques and their impacts towards the properties of natural rubber latex from Hevea Brasiliensis—A comprehensive review related to tire application. Industrial Crops and Products, 181, 114835.

46. Oluseyi, P. O., Babatunde, O. M., & Babatunde, O. A., (2016). Assessment of energy consumption and carbon footprint from the hotel sector within Lagos, Nigeria. Energy and Buildings, 118, 106–113.

47. Onoji, S. E., Iyuke, S. E., Igbafe, A. I., & Nkazi, D. B., (2016). Rubber seed oil: A potential renewable source of biodiesel for sustainable development in sub-Saharan Africa. Energy Conversion and Management, 110, 125–134.

48. Onwukwe, S. I., & Nwakaudu, M. S., (2012). Drilling wastes generation and management approach. International Journal of Environmental Science and Development, 3(3), 252.

49. Peng, X., (2002). Green chemical approaches toward high-quality semiconductor nanocrystals. Chemistry–A European Journal, 8(2), 334–339.

50. Poovaneshvaran, S., Hasan, M. R. M., & Jaya, R. P., (2020). Impacts of recycled crumb rubber powder and natural rubber latex on the modified asphalt rheological behavior, bonding, and resistance to shear. Construction and Building Materials, 234, 117357.

51. Praburanganathan, S., Chithra, S., Divyah, N., Sudharsan, N., Yeddula, B. S. R., & Vigneshwaran, S., (2022). Value-added waste substitution using slag and rubber aggregates in the sustainable and eco-friendly compressed brick production. Revista de la Construcción, 21(1), 5–20.

52. Promlok, D., Sonongbua, K., Wilepsuwan, M., Suteewong, T., & Tangboriboonrat, P., (2022). Hollow natural rubber latex particles as bio-based alternative white pigment for coating applications. Industrial Crops and Products, 188, 115593.

53. Qaidi, S. M., Dinkha, Y. Z., Haido, J. H., Ali, M. H., & Tayeh, B. A., (2021). Engineering properties of sustainable green concrete incorporating eco-friendly aggregate of crumb rubber: A review. Journal of Cleaner Production, 324, 129251.

54. Qambrani, N. A., Rahman, M. M., Won, S., Shim, S., & Ra, C., (2017). Biochar properties and eco-friendly applications for climate change mitigation, waste management, and wastewater treatment: A review. Renewable and Sustainable Energy Reviews, 79, 255–273.

55. Rad, S. J., & Lewis, M. J., (2014). Water utilization, energy utilization and waste water management in the dairy industry: A review. International Journal of Dairy Technology, 67(1), 1–20.

56. Robert, T., Eschig, S., Biemans, T., & Scheifler, F., (2019). Bio-based polyester itaconates as binder resins for UV-curing offset printing inks. Journal of Coatings Technology and Research, 16, 689–697.

57. Sari, M. M., Inoue, T., Septiariva, I. Y., Suryawan, I., Shigeru, K., Harryes, R. K., & Ramadan, B. S., (2022). Identification of face mask waste generation and processing in tourist areas with thermo-chemical process. Archives of Environmental Protection, 48(2).
58. Sarkar, P., & Bhowmick, A. K., (2018). Sustainable rubbers and rubber additives. Journal of Applied Polymer Science, 135(24), 45701.
59. Selvi, S. S., Eminagic, E., Kandur, M. Y., Ozcan, E., Kasavi, C., & Oner, E. T., (2019). Research and production of microbial polymers for food industry. Bioprocessing for Biomolecules Production, 211–238.
60. Shahid, M., & Mohammad, F., (2013). Green chemistry approaches to develop antimicrobial textiles based on sustainable biopolymers— A review. Industrial & Engineering Chemistry Research, 52(15), 5245–5260.
61. Sherriff, B. L., Court, P., Johnston, S., & Stirling, L., (2002). The source of raw materials for Roman pottery from Leptiminus, Tunisia. Geoarchaeology: An International Journal, 17(8), 835–861.
62. Silva, G., Kim, S., Aguilar, R., & Nakamatsu, J., (2020). Natural fibers as reinforcement additives for geopolymers–a review of potential eco-friendly applications to the construction industry. Sustainable Materials and Technologies, 23, e00132.
63. Skinner, J. A., Lewis, K. A., Bardon, K. S., Tucker, P., Catt, J. A., & Chambers, B. J., (1997). An overview of the environmental impact of agriculture in the UK. Journal of Environmental Management, 50(2), 111–128.
64. Tong, T., & Elimelech, M., (2016). The global rise of zero liquid discharge for wastewater management: Drivers, technologies, and future directions. Environmental Science & Technology, 50(13), 6846–6855.
65. Valentini, L., Moore, K. R., & Bediako, M., (2022). Sustainable sourcing of raw materials for construction: From the earth to the moon and beyond. Elements: An International Magazine of Mineralogy, Geochemistry, and Petrology, 18(5), 327–332.
66. Valta, K., Kosanovic, T., Malamis, D., Moustakas, K., & Loizidou, M., (2015). Overview of water usage and wastewater management in the food and beverage industry. Desalination and Water Treatment, 53(12), 3335–3347.
67. Weinert, N., Chiotellis, S., & Seliger, G., (2011). Methodology for planning and operating energy-efficient production systems. CIRP Annals, 60(1), 41–44.

68. Wen, Y., Wang, Y., Zhao, K., & Sumalee, A., (2017). The use of natural rubber latex as a renewable and sustainable modifier of asphalt binder. International Journal of Pavement Engineering, 18(6), 547–559.
69. Zareanshahraki, F., & Mannari, V., (2018). "Green" UV - LED gel nail polishes from bio - based materials. International Journal of Cosmetic Science, 40(6), 555–564.
70. Zhang, Y., Fang, Z. Z., Xia, Y., Huang, Z., Lefler, H., Zhang, T., & Guo, J., (2016). A novel chemical pathway for energy efficient production of Ti metal from upgraded titanium slag. Chemical Engineering Journal, 286, 517–527.

CHAPTER

6 Influence of Additives on Latex Bonding

Contents

6.1. INTRODUCTION

The bonding of latex is of paramount importance in diverse industrial sectors, spanning from sealants and tapes to coatings as well as textiles. The establishment of robust and long-lasting adhesion is a crucial factor in guaranteeing the efficacy and dependability of products that utilize latex as a primary component. The incorporation of additives into latex formulations has a notable influence on both the bonding mechanism and the resultant bond strength. These additives are typically introduced in minute quantities. The utilization of additives in the process of latex bonding acts on multiple objectives. One of the principal purposes is to enhance the adhesive characteristics of the latex, augmenting its capacity to adhere to diverse surfaces. Additives have the ability to alter the surface energy for latex, thereby facilitating wetting and promoting close interaction with the substrate. This is a crucial factor in achieving robust adhesion (Candamano et al., 2021).

In addition, the incorporation of additives has the potential to improve the mechanical characteristics of the bonded interaction. The incorporation of certain additives has the potential to enhance the pliability, resilience, and extensibility within the latex film, thereby yielding adhesive connections that exhibit robustness in the face of mechanical strains, distortions, and ambient factors. Additives play a crucial role in augmenting the mechanical characteristics of the bond, thereby bolstering its durability and sturdiness (Aquino et al., 1995). This, in turn, mitigates the likelihood of separation or failure occurring in the long run. Additives are known to have a significant impact on the optimization of processing as well as the usage of products that are based on latex. The addition of certain substances has the potential to impact the viscosity, flow properties, and curing attributes of latex, thereby facilitating its manageability, coating, or application. This enhances the efficacy and efficiency of the bonding procedure, resulting in consistent and dependable outcomes.

The present chapter offers a thorough investigation into the function of additives in the process of latex bonding. The text encompasses a range of additives that are frequently employed in latex compositions, such as substances that enhance adhesion, agents that facilitate crosslinking, compounds that reduce surface tension, and materials that increase volume. The processes of action, impact on bond strength, as well as impacts on bond longevity and efficiency of each additive, are analyzed. The chapter expounds upon the various factors that impact the efficacy of additives within latex bonding, including parameters related to the composition, features of the substrate, and prevailing environmental conditions. The significance of carefully choosing suitable additives and refining their composition for particular bonding purposes is emphasized. The practical implementation of additives in various industries as well as bonding scenarios is demonstrated through the presentation of case studies. The case studies presented herein offer valuable perspectives on the efficacy of certain additives in enhancing bond endurance, strength, as well as efficiency in practical scenarios, thereby underscoring the advantages of integrating additives within latex bonding procedures. Moreover, the chapter discusses the obstacles and constraints related to the process of additive selection, composition, and compatibility. The text delves into nascent patterns and progressions in additive technology, as well as plausible domains for additional investigation and enhancement (Sani et al., 2021; Qu et al., 2017).

6.2. TYPES OF ADDITIVES IN LATEX BONDING

Adhesion-promoting substances are a class of additives that are specifically formulated to augment the adhesive properties between latex-based contents and diverse substrates. They enhance the development of robust interfacial adhesion through the promotion of molecular interactions along with interlocking processes at the bonding interface.

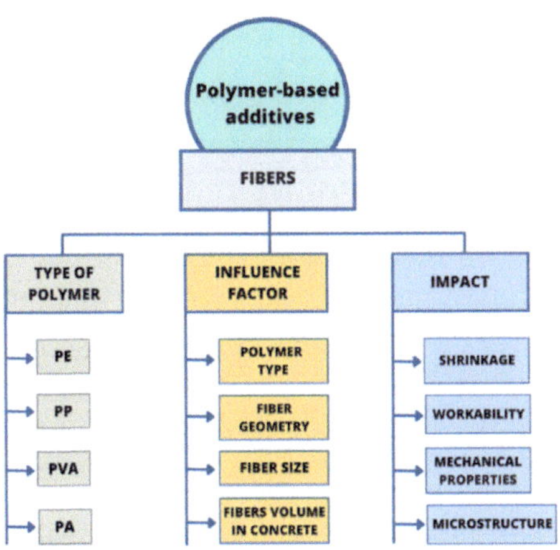

Figure 6.1. The scheme of the type of additives and the changes they cause in latex-based composites.

Source: Weronika Kujawa, Creative Commons License.

The efficacy of adhesion promoters is contingent upon the bonding system as well as the substrate in question, as various mechanisms are employed to facilitate adhesion. The operational mechanism of these substances involves augmenting the wetting capacity of the latex upon the substrate surface, mitigating surface tension, and ameliorating the dispersion of the latex. This facilitates close interaction and enhances the bonding between the latex as well as the substrate (Jutten et al., 1989) (Figure 6.1).

Moreover, the utilization of adhesion promoters can facilitate the establishment of chemical bonds between the latex as well as the substrate. These entities have the ability to undergo chemical reactions with their respective latex as well as the substrate, resulting in the formation of covalent bonds or just specific interactions between molecules. The process of chemical bonding serves to augment the adhesive potency and durability of the bond. Adhesion-promoting substances can serve as coupling agents that connect the latex as well as the substrate, thereby establishing a connection that enhances compatibility as well as interfacial bonding. Enhancement of the mechanically linked mechanism between the latex, as well as the substrate, can lead to the formation of robust and long-lasting bonds (Tran et al., 2022; Pan et al., 2019).

Silanes, titanates, zirconates, as well as organosilanes are among the frequently encountered types of adhesion promoters. Additives are frequently employed in adhesive compositions, coatings, as well as composite materials to augment adhesion to diverse substrates, including plastics, steel, and glass, as well as ceramics (Mavar & Skazlic, 2012).

6.2.1. Crosslinking Agents and their Influence

Crosslinking agents are utilized as additives to facilitate the establishment of chemical bonds within polymer chains present in the latex, resulting in the development of a network with three dimensions architecture. The method of crosslinking serves to augment the potency and durability of the adhesive connection, in addition to the physical attributes of the latex membrane. Crosslinking agents have the ability to undergo a chemical reaction with the functional groups that are present in the latex, including but not limited to carboxyl, hydroxyl, or just amine groups, resulting in the formation of covalent bonds. The formation of covalent bonds among the polymer chains engenders robust chemical interconnections, thereby yielding a stiffer and more enduring configuration (Durif et al., 2015; Nejad et al., 2012).

Tip

When working with latex bonding, carefully select crosslinking agents that are compatible with the latex system to achieve the desired bonding strength and durability.

The incorporation of crosslinks within the latex film enhances its capacity to withstand deformation, stress, as well as environmental stressors. The mechanical characteristics of the bond, including tensile strength, agility, as well as resilience to removing or breaking forces, are improved through its enhancement. Crosslinked latex films demonstrate enhanced resistance to chemicals, thermal endurance, and resilience to moisture or just solvent exposure. The selection of crosslinking agents is contingent upon the particular latex composition and the intended characteristics. Illustrative instances comprise peroxides, azo substances, isocyanates, as well as epoxy resins. The careful choice of a crosslinking agent has paramount importance in order to guarantee the efficacy of the agent within the latex system as well as the requisite bonding specifications (Poovaneshvaran et al., 2020).

The act of crosslinking may be activated through a variety of means, such as the application of heat, exposure to UV radiation, utilization of chemical initiators, or just a combination of these techniques. It is imperative to exercise meticulous control over the ideal crosslinking circumstances, including temperature, time, as well as curing processes, in order to attain the intended crosslink density along with bond properties (Solodkyy et al., 2019).

6.2.2. Surfactants and Their Impact

Surfactants are a class of compounds that exhibit surface activity and are commonly employed in latex compositions to alter the wetting characteristics and enhance the bonding that occurs between latex as well as the underlying surface. The mechanism of action involves the reduction of surface tension within the latex, which facilitates its uniform and efficient spreading on the substrate exterior. The soaking ability of latex may be improved through the use of surfactants, which effectively reduce the degree of contact within the substrate as well as the latex. Enhancing the proximity with the latex as well as the substrate results in improved adhesion and minimizes the occurrence of defects, like insufficient wetting or just air pockets (Schroeder et al., 2010) (Figure 6.2).

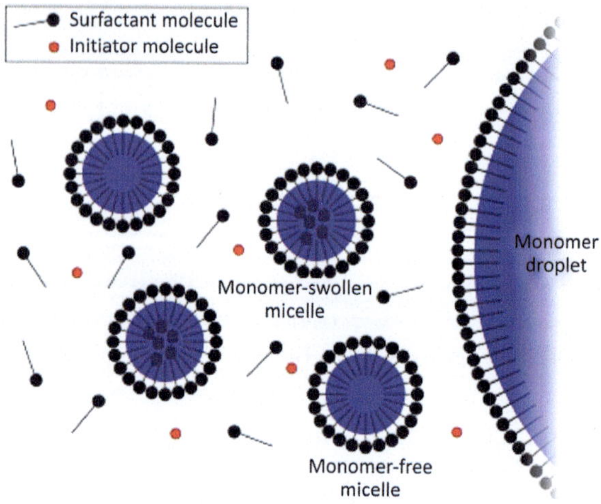

Figure 6.2. Example of non-leaching reactive surfactants for architectural latex binders.

Source: Nathan Noyes, Creative Commons License.

Furthermore, the utilization of surfactants can enhance the resilience of latex scattering through the prevention of particle coagulation as well as aggregation. The stabilizing agents function to decrease the surface energy regarding the latex particles while establishing electrostatic or just steric repulsion, thereby preserving a consistent and stable dispersion. This improves the overall adhesion performance and coating creation of the latex. The choice of surfactants is contingent upon the particular demands of the adhesive system and its congruity with the latex composition. Various categories of surfactants, including anionic, cationic, nonionic, or amorphous surfactants, can be employed depending on the characteristics of the latex as well as the substrate (Geng et al., 2017; Bazmara et al., 2018).

Did You Know?

Surfactants play a crucial role in latex bonding by promoting wetting and adhesion between substrates, enhancing the overall bonding performance.

The chemical makeup of surfactants can exert an impact on the bond that exists among the substrate as well as the latex. Certain surfactants possess functional structures that may interact with the outer layer of the substrate, thereby augmenting the strength of bonds and furnishing supplementary adhesion processes. Surfactants are essential in enhancing wetting acts, strengthening latex dispersions, as well as improving adhesion within latex bonding usages (Berndt & Mancini, 2004).

6.2.3. Fillers and their Effects on Mechanical Properties

Solid particles, commonly referred to as fillers, are incorporated into latex formulations with the aim of enhancing the mechanical characteristics associated with the bonded interactions. The reinforcement of the latex film can be achieved through the enhancement of its durability, toughness, as well as resistance to deformation, thereby leading to an improvement in the general mechanical efficiency of the bond. Incorporating fillers into the latex film can enhance its tensile strength, flexibility, as well as impact resistance. Fillers serve as augmenting agents that offer supplementary structural robustness and rigidity to the adhered interface. Additionally, they have the potential to augment structural integrity and mitigate the effects of creeps or just relaxation on the adhesive bond (Hoy et al., 2023) (Figure 6.3).

Remember

Incorporating fillers into latex bonding formulations can improve mechanical properties such as strength and toughness, providing enhanced performance and durability to bonded assemblies.

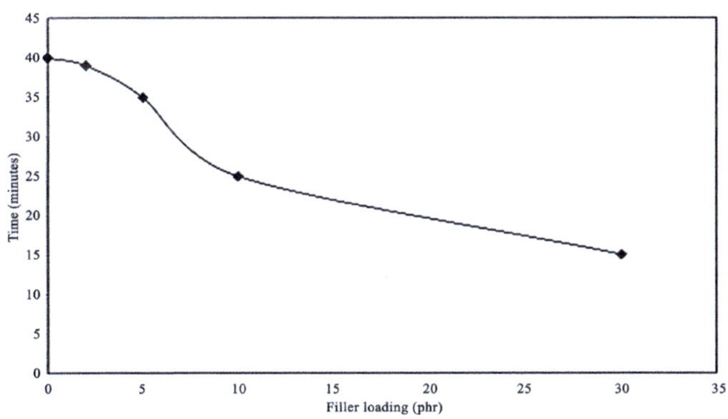

Figure 6.3. Effect of filler loading on the optimum prevulcanization time of prevulcanized natural rubber latex compound.

Source: Manroshan Singh, Creative Commons License.

The selection of filler substance is contingent upon the intended characteristics and the particular bonding context. Inorganic substances along with organic materials are commonly employed as filler substances in the process of latex bonding. Inorganic substances, including silica, carbonate of calcium, talc, along with clay, are frequently utilized, as are organic materials, like fibers made from cellulose or just microspheres. The mechanical attributes of an interface that is bonded can be influenced by the particle size, form, as well as surface features of fillers. Reducing particle size alongside increasing surface area can potentially augment the adhesion and cohesion within the filler as well as the latex matrix, leading to superior mechanical properties (Xie et al., 2022; Zak et al., 2016).

Apart from mechanical augmentation, fillers have the potential to influence various other characteristics associated with the latex film, including its rheological acts, thermal insulation, and barrier characteristics. The modification of viscosity, permeability, as well as drying properties of the latex creation, can significantly impact the bond's effectiveness and simplicity of application. The appropriate dispersion, adhesion, as well as compatibility regarding the filler ingredient with both the latex as well as a substrate should be taken into account to ensure its compatibility. It might be necessary to apply exterior treatment or functioning to the filler particles in order to augment their suitability with the latex matrix as well as improve their cohesive properties (Wennekes et al., 2007; Zhang et al., 2019).

To summarize, the process of selecting and integrating additives into latex bonding has a significant influence on the mechanical characteristics, resilience, and bonding capacity that characterize the bonded interfaces. Adhesion promoters serve for improving wetting as well as interfacial attachment, crosslinking substances facilitate the building and maintaining of bonds, surfactants increase wetting behavior along with stability, as well as fillers solidify the mechanical attributes associated with the bonded interfaces. Comprehending the function and impact of said additives is imperative in the optimization of latex adhesion for particular uses and the attainment of robust, dependable, and high-caliber bonds.

6.3. INFLUENCE OF ADDITIVES ON LATEX BONDING STRENGTH

Diverse additives assume a pivotal function in augmenting the adhesive strength within latex bonding through the utilization of different modes of action.

1. Adhesion promoters are utilized to enhance the adhesive properties of a material by promoting intermolecular interactions as well as chemical bonding within the substrate as well as the latex, thereby improving the durability of the bond.

   The substrate's surface features can be modified, resulting in a decrease in its outermost energy along with an enhancement in its wetting characteristics. This process facilitates improved interfacial interaction with the latex as well as the substrate, resulting in enhanced adhesion. Furthermore, adhesion promoters have the capability to undergo a chemical reaction alongside the functional bands that are present in both the latex as well as the substrate. This reaction results in the formation of ions or covalent bonds, thereby enhancing the overall strength of the bond (Ekinci et al., 2019).

2. Crosslinking agents are utilized to improve the adhesive properties of latex film by inducing the formation of a network that is three-dimensional. The introduction of crosslinking substances within a latex creation facilitates a reaction with the functional groups that are inherent in the latex, thereby leading to the creation of covalent connections within polymer chains. The method of crosslinking enhances the mechanical characteristics of the latex film, including tensile strength, flexibility, as well as deformation resistance, thereby augmenting the bond strength (Baghini et al., 2014).

3. Surfactants are known to enhance the adhesive properties of latex by decreasing its surface tension, thereby facilitating superior wetting as well as spreading upon the outermost layer of the substrate.

   The process facilitates close contact between the latex and the substrate, thereby improving the interfacial attachment. Surfactants have the ability to maintain the dispersed form of latex, thereby preventing the accumulation of particles and promoting uniform coating. This, in turn, leads to an enhancement in the bonding capacity (Shen et al., 2005; Kessel et al., 2008).

4. Fillers are utilized to improve the adhesive properties of latex film by strengthening it. The integration of fillers enhances the mechanical characteristics of the adhered interface, encompassing stiffness, flexibility, and resilience to impact. Fillers serve as

reinforcing agents, imparting supplementary rigidity to the adhesive joint. Additionally, they have the potential to augment the dimensional consistency and bolster the resistance regarding deformation or just creep, thereby fortifying the bond's durability (Akinyemi et al., 2020).

6.3.1. Factors Affecting the Effectiveness of Additives

The efficacy of additives within the bond formation is influenced by multiple factors, and it is imperative to take these factors into account in order to attain maximum bond potency.

1. **Compatibility:** The crucial factor to consider is the integration between the additives and the latex formulations. It is imperative that the additive exhibits compatibility in relation to the latex polymer to facilitate adequate dispersion as well as interaction. The occurrence of incompatibility may result in compromised bond strength due to phase division, poor wetting, or just diminished effectiveness with respect to the additive (Stoeckel et al., 2013).
2. **Formulation Parameters:** The bonding capacity can be considerably affected by the formulation variables, which include the additives concentration, pH, as well as the circumstances of curing. The determination of optimal quantities of additives is necessary to attain the ideal bond efficiency. The responsiveness of additives, specifically crosslinking agents, as well as the bond strength, may be influenced by various factors like pH as well as curing temperatures.
3. **Substrate Characteristics:** The features and inherent qualities of the substrate are of paramount importance in the process of forming bonds. The wetting along with adhesion attributes can be influenced by various factors such as the chemical makeup, porosity, roughness, as well as surface energy that comprise the substrate. The selection of additives ought to be made with careful consideration of the unique properties of the substrate in order to guarantee appropriate interaction as well as adhesion (Sani et al., 2020).
4. **Environmental Conditions:** The bond strength might be influenced by various environmental factors such as temperature, humidity, as well as involvement with chemicals, or just

moisture. Certain additives may demonstrate diverse levels of efficacy in distinct environmental circumstances. The assessment of the consistency and efficacy of additives within the designated application setting is imperative (Šedivka et al., 2015).

6.3.2. Case Studies in Various Applications

Case studies offer tangible illustrations of the impact of additives on the strength of bonds across various applications.

1. The application of adhesives is a common practice in the automotive sector, particularly in the bonding of seals made of rubber for metallic surfaces. Adhesion promoters are typically incorporated into latex-based sealants to enhance their bonding properties. Adhesive promoters are known to augment the strength of the bond by ameliorating the wetting process and catalyzing the chemical bonding of the latex binding as well as the metal base, thereby leading to the formation of sturdy and long-lasting bonds (Li et al., 2022) (Figure 6.4).

Figure 6.4. Image showing the use of adhesion promoters.

Source: Designetics, Creative Commons License.

2. The application of coatings involves the utilization of surfactants within latex-based formulations to enhance the bonding properties with diverse substrates, including plastics and wood. The application of surfactants results in a decrease in the surface pressure of the latex, thereby facilitating its even distribution on

the substrate. As a consequence, there is an enhancement in the wetting process, an increase within the contact area, as well as an improvement in bond strength (Baueregger et al., 2014).

3. The incorporation of fillers is a crucial aspect of composite manufacturing, particularly in the context of composite materials that utilize latex for a matrix. The addition of fillers, like silica or just carbon black, enhances the mechanical characteristics of the composite material, including its bonding strength. The incorporation of fillers into the latex matrix serves to augment the interfacial adhesion and overall mechanical properties of the composite substance (Sikora et al., 2021).

The presented case studies exhibit the noteworthy influence of additives on adhesive strength in diverse scenarios. The appropriate identification and enhancement of additives have the potential to result in more robust, enduring, and superior-quality latex adhesions, thereby guaranteeing the dependability and durability of bonded commodities.

To conclude, the bond strength within latex bonding has a major impact by additives through multiple pathways of behavior. Adhesion promoters facilitate wetting as well as support chemical bonds, crosslinking substances engender a network arrangement to augment mechanical durability, surfactants intensify wetting as well as spreading, while fillers fortify the bonded interfaces. The efficacy of additives can be influenced by various factors, including but not limited to compatibility, manufacturing variables, substrate characteristics, as well as environmental conditions. Comprehending the aforementioned factors along with the significance of additives is imperative in the process of enhancing latex bonding and attaining resilient, enduring, and high-caliber bonds across diverse applications (Buritatun et al., 2020).

6.4. INFLUENCE OF ADDITIVES ON BOND DURABILITY AND PERFORMANCE

The longevity and efficacy of bonded goods are contingent upon the careful consideration of the longevity as well as aging of latex bonds. The utilization of additives within latex bonding has a noteworthy impact on the bond's durability against aging as well as the overall lifespan of the bond.

1. **Adhesion Promoters:** They are capable of augmenting the enduring durability of latex bonds through the facilitation of

robust interfacial adhesion. The enhancement of compatibility across latex along with substrate is known to mitigate the likelihood of bond deterioration over a prolonged period. Adhesion-promoting agents that establish ionic or covalent bonds along with the substrate have the potential to enhance resistance against environmental influences and aging (Ye et al., 2003).

2. **Crosslinking Agents:** They are essential for enhancing the adhesive strength of the latex film by generating a crosslinked structure. The cross-linked configuration of this structure confers improved resistance to deformations, stress, as well as aging of the bond. The cross-linking of latex bonds results in enhanced dimensional stability, greater resistance to creep, as well as decreased vulnerability to environmental variables such as humidity and temperature (Nimpaiboon et al., 2013).
3. **Surfactants:** Surfactants have the potential to enhance the durability of latex connections by facilitating wetting and mitigating the likelihood of delamination or debilitation in the long run. The utilization of these agents is imperative for ensuring sustained intimate contact within the latex as well as the substrate, thereby enhancing the longevity of the bond. Furthermore, surfactants possess the capability to establish a safeguarding shield against humidity or other external elements, thereby conserving the soundness of the bond (Zumbiihl et al., 2007).
4. **Fillers:** They possess the capacity to augment the longevity of latex adhesives through the enhancement of the mechanical characteristics of the adhered interface. The inclusion of fillers has the potential to augment the bond's capacity to withstand impact, deformation, as well as stress. Moreover, certain fillers have the ability to function as strengthening agents, thereby enhancing their durability against degradation and preserving their adhesive efficacy for a prolonged duration (Machotová et al., 2020).

6.4.1. Resistance to Environmental Factors

The incorporation of additives within latex bonding has been found to exert a substantial impact on the bond's capacity to withstand environmental stressors and chemical agents, both of which are frequently encountered by bonded articles.

1. The incorporation of specific additives, like adhesion promoters or just crosslinking agents with moisture-resistant properties, can enhance the humidity resistance of the bond. It is possible to generate hydrophobic barriers or amplify the hydrophobic property within the latex film, thereby diminishing the uptake of moisture as well as forestalling bond deterioration due to swelling or just weakening (Centauro et al., 2017).
2. The bond's ability to withstand high temperatures can be improved by incorporating crosslinking agents along with heat-resistant additives. The cross-linked configuration confers thermal stability, thereby enabling the bond to preserve its rigidity and strength even in circumstances of elevated temperature. The incorporation of heat-resistant ingredients has the potential to enhance the bond's capacity to withstand softening or just deformation when exposed to high temperatures (Benali & Ghomari, 2017).
3. Chemical resistance may be improved by incorporating certain additives, like chemical-resistant attachment promoters or just fillers, which can augment the bond's ability to withstand chemical exposure. Barrier properties can be exhibited by them, thereby preventing chemical breakdown or penetration through the bond. Furthermore, it is commonly observed that crosslinked latex films exhibit heightened chemical resistance, thereby promoting the longevity of the adhesive bond in chemically harsh settings (Mousavi & Dehestani, 2022).

6.4.2. Case Studies in Different Conditions

Numerous case studies have exhibited the influence of additives on the endurance of bonds under diverse circumstances.

1. Latex-based glues with UV-resistant contaminants have been utilized in the construction sector for outdoor connecting applications. The aforementioned additives serve the purpose of safeguarding the bond against the harmful effects of UV radiation, thereby impeding bond deterioration and sustaining its long-lasting robustness and efficacy (Mohammadi et al., 2014).
2. Latex coatings containing moisture-resistant contaminants have been utilized in applications such as bathrooms or kitchens, where moisture-rich circumstances are present. The incorporation of these additives enhances the adhesive's capacity to withstand

moisture as well as humidity, thereby mitigating the likelihood of bond failure and preserving the coating's structural soundness throughout its lifespan (Yang et al., 2022).

3. Latex-based adhesives containing chemical-resistant contaminants are employed in industrial or chemical processing settings due to their chemical resistance properties. The incorporation of these additives confers augmented resilience against exposure to chemicals, thereby ensuring the bond's endurance and efficacy in conditions that are corrosive or just chemically aggressive (Raza et al., 2022).

The aforementioned case studies serve to underscore the notable impact of additives on the longevity of bonding in diverse chemical as well as environmental circumstances. The judicious choice and integration of additives can enhance the enduring stability, resilience to environmental stressors, and chemical susceptibility of latex adhesives, thereby ensuring dependable and long-lasting functionality across a range of applications (Marcinko et al., 2020). To summarize, the incorporation of additives within latex bonding is of significant importance in augmenting the durability and overall performance of the bond. They facilitate the attainment of prolonged stability and durability of bonds, augment resilience against external elements such as humidity and temperature, and reinforce resistance against chemical contact. The efficacy of additives in enhancing bond durability under diverse conditions is demonstrated through the analysis of case studies. The longevity, as well as durability of latex bonds, may be ensured by carefully considering the particular environmental issues and selecting appropriate additives.

6.5. OPTIMIZATION AND SELECTION OF ADDITIVES

The selection of suitable additives plays a pivotal role in attaining optimal bonding characteristics for particular bonding applications. Multiple factors need to be considered:

1. Comprehending the particular bonding requisites is imperative for successful application. The choice of contaminants will be guided by factors like the envisioned bond rigidity, adaptability, resilience, as well as environmental resistance. In instances where a strong bond is of utmost importance, crosslinking agents can be selected. In situations where malleability is deemed significant, the implementation of plasticizers or just elastomers might get deemed appropriate (Simatupang et al., 1994).

2. The assessment of substrate compatibility is imperative, whereby the compatibility within the substrate and additives must be thoroughly examined. The chemical compositions along with the surface properties of diverse substrates necessitate the selection of appropriate additives for optimal adhesion as well as compatibility. The utilization of adhesion supporters that are congruous with the material of the substrate has the potential to augment interfacial bonding and ultimately enhance the entire bond performance (Jo et al., 2014).
3. The compatibility of contaminants with latex manufacturing is crucial to ensure appropriate dispersion, stability, as well as curing properties. Incompatibility concerns may result in the amalgamation, phase segregation, or diminished efficacy of the additives, thereby jeopardizing the bonding characteristics. The selection of additives involves crucial stages such as compatibility testing as well as formulation optimization (Huo et al., 2016).

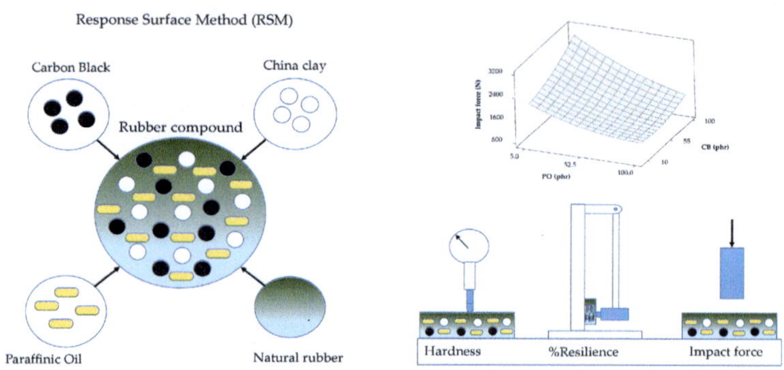

Figure 6.5. Optimization and selection of additives based on response surface method (RSM).

Source: Satta Srewaradachpisal, Creative Commons License.

4. The assessment of environmental conditions is crucial in determining the expected operational circumstances as well as environmental variables that may impact the system's performance. Factors such as moisture, temperature, contact with chemicals, or ultraviolet (UV) rays should be taken into account during the evaluation process. The incorporation of additives that

confer resistance to environmental variables, like UV stabilizers as well as moisture-resistant substances, can effectively preserve the structural and functional integrity of the bond within demanding environmental conditions (Fu et al., 1995) (Figure 6.5).

6.5.1. Strategies for Optimizing Additive Formulation

The optimization of the additive formulation seems a crucial aspect in attaining the intended bonding characteristics. Numerous tactics can be utilized.

1. Optimization of manufacturing involves the adjustment of additive concentration throughout the latex composition, which can exert a substantial influence on the properties of bonding. To attain the ideal combination of properties, it is imperative to conduct systematic tests and assessments to determine the ideal amount of each additive (Contamine & Si Larbi, 2016).
2. The attainment of desired bond characteristics may require an assortment of additives in numerous instances. The overall bonding performance may be enhanced by combining various additives, including adhesion supporters, crosslinking agents, as well as surfactants, to generate synergistic benefits. It is imperative to conduct a thorough evaluation of the compatibility among additives as well as potential interactions (Zárybnická et al., 2023).
3. The consideration of additive interactions is imperative in the formulation process, particularly in relation to the relationships between various additives. Certain additives may exhibit synergistic or opposing effects upon being combined. Conducting compatibility as well as stability assessments can facilitate the detection of conceivable interactions and steer the choice of additives (Xu et al., 2016).
4. The improvement of the method involves the examination of manufacturing conditions such as temperature, mixing duration, and curing variables, which have the potential to influence the efficacy of additives. Maximizing the efficacy of additives along with ensuring reliable and easily reproducible bond characteristics can be achieved by enhancing the conditions used for processing (John et al., 2023).

6.5.2. Challenges in Additive Formulation

The process of selecting and formulating additives within latex bonding can come with various challenges along with limitations.

1. The attainment of compatibility among additives, the latex composition, as well as the substrate may represent a challenging task. The occurrence of incompatibility may result in inadequate dispersion, diminished efficacy of additives, or just instability within the latex formulations. In order to tackle these challenges, it is imperative to conduct compatibility testing as well as optimization (Kantarci et al., 2021).
2. Performance trade-offs exist when considering various additives as they may offer distinct advantages but also have certain drawbacks. As an illustration, specific adhesion promoters have the potential to enhance the strength of bonding; however, they may also have an adverse impact on the flexibility of the material. Achieving a balance between the desired characteristics and mitigating potential trade-offs is a crucial aspect of the selection and formulation of additives (Sukumar et al., 2020).
3. Cost implications must be taken into account when considering the use of certain additives, particularly those with different characteristics or specialized features, as they may be relatively pricey. Optimizing the whole formulation requires the balancing of performance requirements and cost considerations (Nazerian et al., 2018).
4. Environmental as well as regulatory factors should be taken into account when considering the use of certain additives, as they may have potential implications in these areas. The evaluation of the potential effects of additives on safety, well-being, and environmental factors is a crucial consideration. The utilization of eco-friendly or low-toxicity contaminants is frequently deemed advantageous.

To succinctly encapsulate the matter, the optimization as well as a selection of contaminants in the context of latex bonding necessitates a meticulous assessment of bonding requisites, substrate congruity, ambient circumstances, and formulation methodologies.

The attainment of desired bond characteristics can be facilitated through the implementation of key strategies such as manufacturing efficiency, combining various additives, as well as procedure optimization. The effective use of additives within latex bonding requires addressing various challenges, including compatibility difficulties, efficiency trade-offs, costs, as well as environmental factors (Pagnoux et al., 1998).

Activity 6.1

A manufacturer is developing a new latex bonding adhesive for automotive applications. They want to enhance the bond strength and durability of the adhesive under high-temperature conditions. Which types of additives would you recommend incorporating into the formulation, and how would they influence the bonding performance?

6.6. SUMMARY

The chapter titled "Influence of Additives on Latex Bonding" explores the crucial function of additives in augmenting bond strength, longevity, and efficacy in latex bonding scenarios. The chapter delves into various categories of additives that are frequently employed, such as adhesion promoters, cross-linking intermediaries, surfactants, as well as fillers. The introductory section of the chapter delves into the processes of action pertaining to every kind of additive. Adhesion promoters enhance the strength of bonding by means of improved wetting, chemical interactions, as well as intermolecular bonding. The utilization of crosslinking agents results in the formation of a three-dimensional network, thereby enhancing the rigidity and stability of the bond. Surfactants enhance the wetting and adhesion characteristics of a substance by decreasing its surface tension, whereas fillers strengthen the mechanical attributes of the bonded interfaces. The efficacy of additives in promoting bond formation is comprehensively investigated, taking into account various influencing factors. Detailed discussions are presented regarding various considerations, including integration with the substrate as well as latex, manufacturing parameters, as well as environmental conditions. The article showcases case studies that demonstrate the influence of additives on the bonding efficacy in diverse applications including adhesives, coatings, as well as composites. The chapter places particular emphasis on maximizing efficiency and the selection of contaminants for bonding applications that are specific in nature. The article emphasizes the importance of carefully selecting additives according to bonding requirements, taking into account factors such as bond rigidity, adaptability, and durability. The present discourse delves into the various techniques that can be employed to enhance the additive formulation procedure with the aim of achieving the desired bond characteristics. These techniques include manufacturing optimization, additives combination, as well as process optimization.

The present study addresses the challenges and limitations associated with the selection and formulation of additives. These include compatibility problems, efficiency trade-offs, expenses, and environmental variables. The chapter highlights the significance of tackling these challenges to guarantee the efficacious execution of additive techniques in latex bonding. The chapter concludes by offering a comprehensive comprehension of the impact of contaminants on the bonding for latex. Industries can attain robust, enduring, and high-performing latex bonds by carefully selecting and integrating appropriate additives, and maximizing their formulation while taking into account particular bonding prerequisites. The comprehension acquired from this chapter confers the ability to investigators, engineers, as well as professionals to make judicious determinations in the selection, composition, and implementation of additives for the purpose of augmenting latex bonding.

REVIEW QUESTIONS

1. What are the main types of additives used in latex bonding?
2. How do crosslinking agents influence latex bonding?
3. What is the impact of surfactants on latex bonding?
4. How do fillers affect the mechanical properties of latex bonding?
5. What factors can affect the effectiveness of additives in latex bonding?
6. Provide examples of case studies showcasing the influence of additives in various applications.
7. How do additives contribute to bond durability and performance in different environmental conditions?

REFERENCES

1. Akinyemi, B. A., Ojediran, J., Olawale, O., & Ayanwola, S., (2020). Efficacy of expanded polystyrene as fine aggregate in cement mortars modified with latex paint as an alternative to polymer admixture. Journal of the Mechanical Behavior of Materials, 29(1), 163–168.
2. Aquino, M. J., Li, Z., & Shah, S. P., (1995). Mechanical properties of the aggregate and cement interface. Advanced Cement Based Materials, 2(6), 211–223.
3. Baghini, M. S., Ismail, A., Karim, M. R., Shokri, F., & Firoozi, A. A., (2014). Effect of styrene–butadiene copolymer latex on properties and durability of road base stabilized with Portland cement additive. Construction and Building Materials, 68, 740–749.
4. Baueregger, S., Perello, M., & Plank, J., (2014). Influence of anti-caking agent kaolin on film formation of ethylene–vinylacetate and carboxylated styrene–butadiene latex polymers. Cement and Concrete Research, 58, 112–120.
5. Bazmara, B., Tahersima, M., & Behravan, A., (2018). Influence of thermoplastic polyurethane and synthesized polyurethane additive in performance of asphalt pavements. Construction and Building Materials, 166, 1–11.
6. Benali, Y., & Ghomari, F., (2017). Latex influence on the mechanical behavior and durability of cementitious materials. Journal of Adhesion Science and Technology, 31(3), 219–241.
7. Berndt, M. L., & Mancini, C. E., (2004). Tensile tests on plain and fiber reinforced geothermal cements. Journal of Materials Science, 39, 1727–1733.
8. Buritatun, A., Takaikaew, T., Horpibulsuk, S., Udomchai, A., Hoy, M., Vichitcholchai, N., & Arulrajah, A., (2020). Mechanical strength improvement of cement-stabilized soil using natural rubber latex for pavement base applications. Journal of Materials in Civil Engineering, 32(12), 04020372.
9. Candamano, S., Crea, F., Coppola, L., De Luca, P., & Coffetti, D., (2021). Influence of acrylic latex and pre-treated hemp fibers on cement-based mortar properties. Construction and Building Materials, 273, 121720.
10. Centauro, I., Cantisani, E., Grandin, C., Salvini, A., & Vettori, S., (2017). The influence of natural organic materials on the properties of traditional lime-based mortars. International Journal of Architectural Heritage, 11(5), 670–684.

11. Contamine, R., & Si Larbi, A., (2016). Development of a textile reinforced concrete (TRC) to retrofit reinforced concrete structures. European Journal of Environmental and Civil Engineering, 20(6), 626–642.
12. Durif, C., Frömder, C., Affolter, C., Lippmann, W., & Graule, T., (2015). Aquacasting—A new shaping concept for water based reactive tape casting. Journal of the European Ceramic Society, 35(13), 3633–3640.
13. Ekinci, E., Türkmen, İ., Kantarci, F., & Karakoç, M. B., (2019). The improvement of mechanical, physical and durability characteristics of volcanic tuff based geopolymer concrete by using nano silica, micro silica, and styrene-butadiene latex additives at different ratios. Construction and Building Materials, 201, 257–267.
14. Fu, Y., Gu, P., Xie, P., & Beaudoin, J. J., (1995). Effect of chemical admixtures on the expansion of shrinkage-compensating cement containing a pre-hydrated high alumina cement—Based expansive additive. Cement and Concrete Research, 25(1), 29–38.
15. Geng, S., Shah, F. U., Liu, P., Antzutkin, O. N., & Oksman, K., (2017). Plasticizing and crosslinking effects of borate additives on the structure and properties of poly (vinyl acetate). RSC Advances, 7(13), 7483–7491.
16. Hoy, M., Tran, N. Q., Suddeepong, A., Horpibulsuk, S., Mobkrathok, M., Chinkulkijniwat, A., & Arulrajah, A., (2023). Improved fatigue properties of cement-stabilized recycled materials–Lateritic soil using natural rubber latex for sustainable pavement applications. Transportation Geotechnics, 40, 100959.
17. Huo, N. B., & Zhao, C. H., (2016). Preparation of EVA/PVA/PU Multi-component water-based adhesive. In: 2016 6th International Conference on Applied Science, Engineering and Technology (pp. 269–274). Atlantis Press.
18. Jo, B. W., Chakraborty, S., & Yoon, K. W., (2014). A hypothetical model based on effectiveness of combined alkali and polymer latex modified jute fiber in controlling the setting and hydration behavior of cement. Construction and Building Materials, 68, 1–9.
19. John, S. K., Nadir, Y., Cascardi, A., Arif, M. M., & Girija, K., (2023). Effect of addition of nanoclay and SBR latex on fly ash-slag geopolymer mortar. Journal of Building Engineering, 105875.
20. Jutten, J. J., Guillot, D., & Parcevaux, P. A., (1989). Relationship between cement slurry composition, mechanical properties, and cement-bond-log output. SPE Production Engineering, 4(01), 75–82.
21. Kantarci, F., Türkmen, İ., & Ekinci, E., (2021). Improving elevated temperature performance of geopolymer concrete utilizing nano-silica,

micro-silica, and styrene-butadiene latex. Construction and Building Materials, 286, 122980.

22. Kessel, N., Illsley, D. R., & Keddie, J. L., (2008). The diacetone acrylamide crosslinking reaction and its influence on the film formation of an acrylic latex. Journal of Coatings Technology and Research, 5, 285–297.
23. Li, W., Cao, M., & Chang, J., (2022). Optimized formulation of waterproof coating based on polymer and sodium silicate-activated GGBS using box Behnken design-selection of additives. Journal of Building Engineering, 61, 105220.
24. Machotová, J., Kalendová, A., Voleská, M., Steinerová, D., Pejchalová, M., Knotek, P., & Zarybnicka, L., (2020). Waterborne hygienic coatings based on self-crosslinking acrylic latex with embedded inorganic nanoparticles: A comparison of nanostructured ZnO and MgO as antibacterial additives. Progress in Organic Coatings, 147, 105704.
25. Marcinko, J. J., Lenges, C., Kim, K., & Harvey, M., (2020). Nuvolve engineered polysaccharides as performance-enhancing additives for latex wood adhesives. Forest Products Journal, 70(3), 293–301.
26. Mavar, K., & Skazlic, M., (2012). Influence of construction technology on the adhesion of remedial concrete. Gradevinar, 64(7), 545–552.
27. Mohammadi, M., Moghtadaei, R. M., & Samani, N. A., (2014). Influence of silica fume and metakaolin with two different types of interfacial adhesives on the bond strength of repaired concrete. Construction and Building Materials, 51, 141–150.
28. Mousavi, S. S., & Dehestani, M., (2022). Influence of latex and vinyl disposable gloves as recycled fibers in 3D printing sustainable mortars. Sustainability, 14(16), 9908.
29. Nazerian, M., Nanaii, H. A., & Gargarii, R. M., (2018). Silica (SiO_2) content on mechanical properties of cement-bonded particleboard manufactured from lignocellulosic materials. Drvna Industrija, 69(4), 317–328.
30. Nejad, F. M., Azarhoosh, A. R., Hamedi, G. H., & Azarhoosh, M. J., (2012). Influence of using nonmaterial to reduce the moisture susceptibility of hot mix asphalt. Construction and Building Materials, 31, 384–388.
31. Nimpaiboon, A., Amnuaypornsri, S., & Sakdapipanich, J., (2013). Influence of gel content on the physical properties of unfilled and carbon black filled natural rubber vulcanizates. Polymer Testing, 32(6), 1135–1144.
32. Pagnoux, C., Chartier, T., Granja, M. D. F., Doreau, F., Ferreira, J. M., & Baumard, J. F., (1998). Aqueous suspensions for tape-casting based on acrylic binders. Journal of the European Ceramic Society, 18(3), 241–247.

33. Pan, C., Liang, D., Mo, L., Riara, M., & Lin, J., (2019). Influence of different modifiers on bonding strength and rheological performance of bitumen emulsion. Materials, 12(15), 2414.
34. Poovaneshvaran, S., Hasan, M. R. M., & Jaya, R. P., (2020). Impacts of recycled crumb rubber powder and natural rubber latex on the modified asphalt rheological behavior, bonding, and resistance to shear. Construction and Building Materials, 234, 117357.
35. Qu, X., & Zhao, X., (2017). Influence of SBR latex and HPMC on the cement hydration at early age. Case Studies in Construction Materials, 6, 213–218.
36. Raza, A., Khan, I., Tufail, R. F., Frankovska, J., Mushtaq, M. U., Salmi, A., & Javed, M. F., (2022). Evaluation of moisture damage potential in hot mix asphalt using polymeric aggregate treatment. Materials, 15(15), 5437.
37. Sani, A., Mohd, H. M. R., Shariff, K. A., Jamshidi, A., Ibrahim, A. H., & Poovaneshvaran, S., (2020). Engineering and microscopic characteristics of natural rubber latex modified binders incorporating silane additive. International Journal of Pavement Engineering, 21(14), 1874–1883.
38. Sani, A., Shariff, K. A., Hasan, M. R. M., Ando, T., & Imai, H., (2021). Behavioral interface-bonding and chemical characterization of silane and wax based additives on latex modified asphalt binders. International Journal of Adhesion and Adhesives, 106, 102822.
39. Schroeder, W. F., Liu, Y., Tomba, J. P., Soleimani, M., Lau, W., & Winnik, M. A., (2010). Influence of ethylene glycol and propylene glycol on polymer diffusion in poly (butyl acrylate-co-methyl methacrylate) latex films. The Journal of Physical Chemistry B, 114(9), 3085–3094.
40. Šedivka, P., Bomba, J., Böhm, M., & Boška, P., (2015). Influence of temperature on the strength of bonded joints. BioResources, 10(3), 3999–4010.
41. Shen, Q., Chen, Y., Wei, H., Zhao, Y., Wang, D., & Xu, D., (2005). Suspension effect of poly (styrene-ran-methacrylic acid) latex particles on crystal growth of calcium carbonate. Crystal Growth & Design, 5(4), 1387–1391.
42. Sikora, P., Chougan, M., Cuevas, K., Liebscher, M., Mechtcherine, V., Ghaffar, S. H., & Stephan, D., (2021). The effects of nano-and micro-sized additives on 3D printable cementitious and alkali-activated composites: A review. Applied Nanoscience, 1–19.
43. Simatupang, M. H., Schmitt, U., & Kasim, A., (1994). Wood extractives of rubberwood (Hevea Brasiliensis) and their influences on the setting of the inorganic binder in gypsum-bonded particleboards. Journal of Tropical Forest Science, 269–285.

44. Solodkyy, S. J., Novytskyi, Y. L., Topylko, N. I., & Turba, Y. V., (2019, December). Research of influence of polymer additives-stabilizers on physical-mechanical indicators and microstructure of cement ground. In: IOP Conference Series: Materials Science and Engineering (Vol. 708, No. 1, p. 012107). IOP Publishing.
45. Stoeckel, F., Konnerth, J., & Gindl-Altmutter, W., (2013). Mechanical properties of adhesives for bonding wood—A review. International Journal of Adhesion and Adhesives, 45, 32–41.
46. Sukumar, T., Varghese, J., Bhargavan, S., Jayasree, P., Suvekbala, V., Alaganandam, K., & Ragupathy, L., (2020). Cytotoxicity of formulated graphene and its natural rubber nanocomposite thin film in human vaginal epithelial cells: An influence of noncovalent interaction. ACS Biomaterials Science & Engineering, 6(4), 2007–2019.
47. Tran, N. Q., Hoy, M., Suddeepong, A., Horpibulsuk, S., Kantathum, K., & Arulrajah, A., (2022). Improved mechanical and microstructure of cement-stabilized lateritic soil using recycled materials replacement and natural rubber latex for pavement applications. Construction and Building Materials, 347, 128547.
48. Wennekes, W. B., Datta, R. N., & Noordermeer, J. W. M., (2007). Mechanistic investigations into the adhesion between RFL-treated cords and rubber. Part II: The influence of the vinyl-pyridine content of the RFL-latex. Rubber Chemistry and Technology, 80(4), 565–579.
49. Xie, H., Li, C., & Wang, Q., (2022). A critical review on performance and phase separation of thermosetting epoxy asphalt binders and bond coats. Construction and Building Materials, 326, 126792.
50. Xu, Y., Kuang, Y., Salminen, P. J., & Chen, G., (2016). The influence of nano-fibrillated cellulose as a coating component in paper coating. BioResources, 11(2), 4342–4352.
51. Yang, Y., Ma, L., Wang, H., Jia, W., Zhu, J., Wang, J., & Yang, S., (2022). A novel water-based lubricating additive of GO@ PTFE: Superior tribological performances from the synergistic effect. Tribology International, 169, 107485.
52. Ye, X., Wu, J., Oh, J. K., Winnik, M. A., & Wu, C., (2003). Effect of pluronic surfactants on the polymer diffusion rate in poly (butyl methacrylate) latex films. Macromolecules, 36(23), 8886–8889.
53. Zak, P., Ashour, T., Korjenic, A., Korjenic, S., & Wu, W., (2016). The influence of natural reinforcement fibers, gypsum, and cement on compressive strength of earth bricks materials. Construction and Building Materials, 106, 179–188.

54. Zárybnická, L., Pokorný, J., Machotová, J., Ševčík, R., Šál, J., & Viani, A., (2023). Study of keto-hydrazide crosslinking effect in acrylic latex applied to Portland cements with respect to physical properties. Construction and Building Materials, 375, 130897.
55. Zhang, X., Li, G., & Song, Z., (2019). Influence of styrene-acrylic copolymer latex on the mechanical properties and microstructure of Portland cement/Calcium aluminate cement/Gypsum cementitious mortar. Construction and Building Materials, 227, 116666.
56. Zumbiihl, S., Attanasio, F., Scherrer, N. C., Müller, W., Fenners, N., & Caseri, W., (2007). Solvent action on dispersion paint systems and the influence on the morphology–changes and destruction. In: Modern Paints Uncovered: Proceedings from the Modern Paints Uncovered Symposium (p. 257). Getty Publications.

CHAPTER

7 Applications of Latexes

Contents

7.1. INTRODUCTION

The versatile characteristics and capabilities of latex materials have resulted in their widespread utilization in various industrial applications. These applications span diverse sectors and underscore the importance of latex materials in various industries. Latex is a crucial constituent in the production of automobile tires within the automotive sector. Incorporating latex binders in tire manufacturing is known to augment the mechanical properties of tires, thereby improving their performance and longevity. This is attributed to the reinforcement of the tire's strength and elasticity, which leads to enhanced wear resistance, traction, and security in general while driving. Latex gloves are extensively utilized in medical and industrial environments due to their exceptional barrier properties, flexibility, and tactile sensitivity. The indispensability of gloves in hospitals, research facilities, and other fields necessitating hand protection is attributed to their efficacy in safeguarding against physiological and chemical hazards.

Latex paints are widely employed in the fields of architecture and decoration. The paints mentioned above present benefits such as facile application, minimal olfactory impact, rapid drying, and long-lastingness. Latex paints are commonly preferred for various surfaces due to their extensive color selection, superior adhesion, and ability to resist cracking and peeling. Utilizing latex binders is advantageous for industrial coatings as it aids in safeguarding metallic substances, wood coatings, and anti-corrosion coatings. The capacity of latex binders to form a film guarantees the adhesion, durability, and resilience to environmental stressors, thereby conferring enduring safeguarding to the covered substrates.

Latex coatings are utilized within the pulp and paper sector to improve surface characteristics. Paper additives enhance the surface characteristics of paper, such as its printability, smoothness, and water resistance, thereby rendering it appropriate for applications that necessitate superior printing quality and longevity, such as packaging and labeling (Pichot et al., 2013; Avramenko et al., 2011).

Moreover, latex-based materials exhibit potential and promise in specialized applications. Nanofabrication and optoelectronics are among the fields where they are increasingly being utilized as scientific instruments. Latex materials possess potential uses in systems for delivering drugs, transplantation, and diagnostics within biomedical technology (Peula-García et al., 2002). Ultimately, the industrial utilization of latex materials is evident in diverse fields, demonstrating their multifaceted nature and significance. Latex materials are widely used in various products and applications, ranging from auto tires and rubber gloves to paints made from latex and industrial coatings, owing to their ability to enhance performance, durability, and safety. Due to their versatility and capacity for continued investigation, they constitute a significant element in numerous sectors (McDonald & Devon, 2002).

7.2. INDUSTRIAL APPLICATIONS

According to the literature, latexes are typically not utilized in their pure form but are blended with diverse constituents to customize them for their designated uses.

Latexes are modified by adding pigments and thickening agents to correspondingly alter their optical and rheological characteristics. The primary application of latex is not as a standalone product but rather as an enhancer of the functional characteristics of the final commercial item on which it is administered.

Using latex coatings on industrial equipment offers resistance to chemicals on exposed surfaces. Additionally, the application of latex binders to fabrics provides dimensional stability. Typically, commercial latexes are manufactured in substantial quantities for trade (with an aggregate production of no less than eight million tons of dry latex per annum), and advancements in product quality are primarily attained through evaluating cost-effectiveness with performance (Paques & Macedo, 2006).

7.2.1. Vehicle Tires

The majority of latex production worldwide is attributed to the employment of natural rubber latexes in the manufacturing of tires. Natural rubber is obtained from the rubber plant and undergoes a process of coagulation and drying before being transported to the tire manufacturing facility.

It is combined with artificial rubber to enhance its chemical durability. Adding carbon black to the tire serves the dual purpose of providing its distinctive black hue, effectively concealing dirt, and enhancing its resistance to abrasion and ultraviolet radiation.

Fillers in the tire composition can substitute the relatively expensive rubber material or enhance the mechanical characteristics (Le Roux et al., 2000) (Figure 7.1).

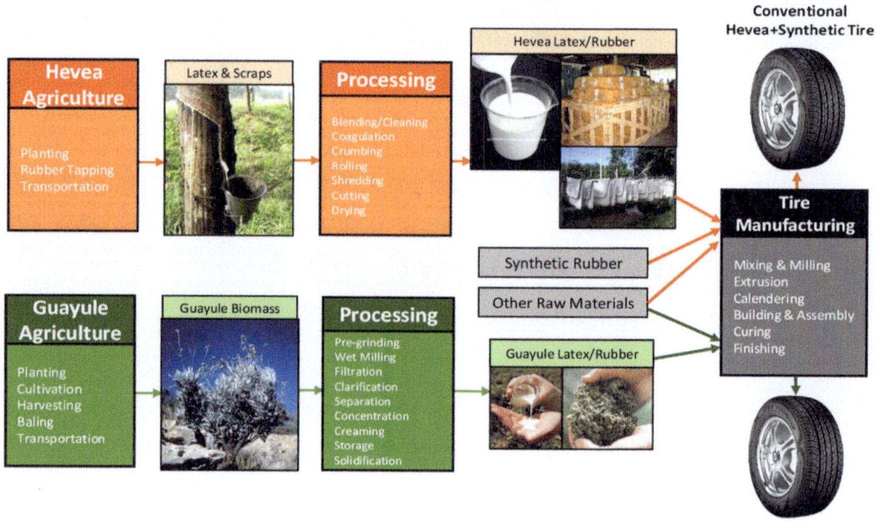

Figure 7.1. Natural latex and synthetic latex for tire manufacturing.

Source: D. Rasutis, Creative Commons License.

The extrusion process is employed to shape the rubber into the desired configuration of the traverse and sidewall. The plies traverse a tire are created by coating fabric cords with rubber, while the belts that run across the tread are formed by rubber-coated steel wires, which is why they are commonly referred to as steel-belted radial tires. The incorporation of plies and belts into the structure of the tire serves to enhance its strength and promote dimensional stability. The constituents of the tire are consolidated onto a cylindrical structure to produce a tire that is in its raw or unvulcanized state. After its production, the green tire is subjected to a vulcanization process by applying sulfur at a temperature of 175 °C within a two-piece mold. This process facilitates the crosslinking of the tire. The optimal crosslink density must be achieved to confer stiffness to the tire while avoiding excessive crosslinking that could render the tire fragile (Sharma, 2013).

7.2.2. Latex Gloves

Natural rubber (NR) latexes are the predominant material used to produce latex gloves utilized by doctors, nurses, emergency medical workers, police officers, and fire personnel. The gloves are produced through a rubber-dipping technique, in which a mold or form, typically resembling a hand with a forearm, is immersed in a solution containing natural rubber latex. A slender latex polymer coating adheres to the mold upon removal from the solution. Subsequently, the assembly is subjected to high temperatures to desiccate and shape the glove, culminating in the detachment of the glove from the mold (Turjanmaa, 1987) (Figure 7.2).

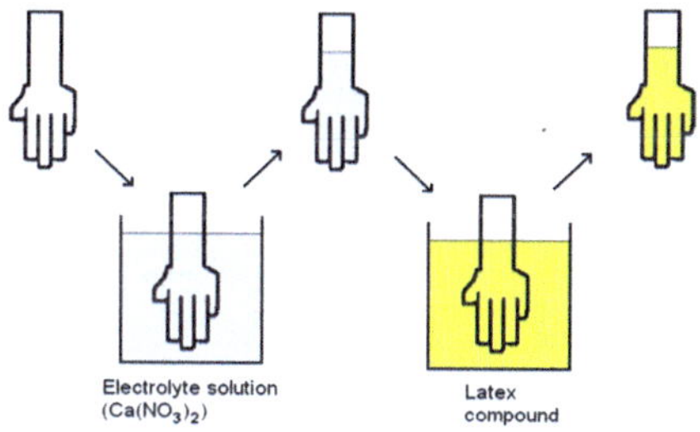

Figure 7.2. Latex gloves dipping process.

Source: Alex Routh, Creative Commons License.

The dipping process exhibits several variations that encompass diverse aspects such as the manufacturing process, which can be batch-based for thicker gloves or persistent for thinner disposable gloves, the makeup of the latex bath, the utilization of coagulant solutions, the conditions for drying and vulcanization, the cleaning of gloves through a leaching process to extract obtainable natural rubber latex proteins to mitigate allergenicity, heat, the use of anti-tack material, chlorination, among others. One possible method involves immersing the form in a coagulant solution, such as calcium nitrate or carbonate, followed by partial drying and dipping into the latex bath. Applying the coagulant onto the surface would induce flocculation of the latex, leading to the formation of a polymer film on the substrate. The control of latex film thickness and subsequent glove production can be achieved through the manipulation of various factors, including immersion duration of the form in the latex bath, the solids content of the latex, viscosity of the latex, the amount of the coagulant, and incorporation of additives (Tarlo et al., 1994).

7.2.3. Latex Paints

Latex or water-based paints are frequently used for architectural embellishments in interior and exterior applications when painting houses and buildings. Latex paints have three primary constituents: binding agents, colors, and water. The binders refer to the latex particles which, upon desiccation, generate a polymer film on the surface. The polymerization process in latex paint is responsible for its robustness and resilience. The latex particles have been engineered to exhibit a glass transition temperature (Tg) of approximately ambient temperature, thereby facilitating the formation of a continuous safeguarding polymer layer through film formation within a relatively short period (i.e., a few days) following water removal through drying. The polymer exhibits sufficient firmness to prevent tackiness and the accumulation of dirt or dust (Chang et al., 1999) (Figures 7.3 & 7.4).

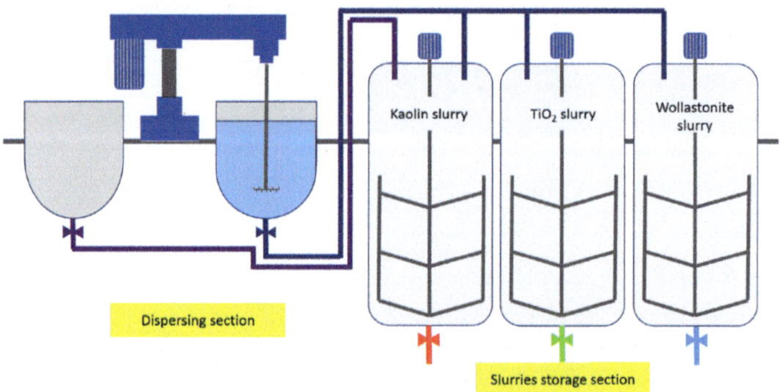

Figure 7.3. Slurry process – stage I – dispersing and slurries storage in the tanks.

Source: Artur Palasz, Creative Commons License.

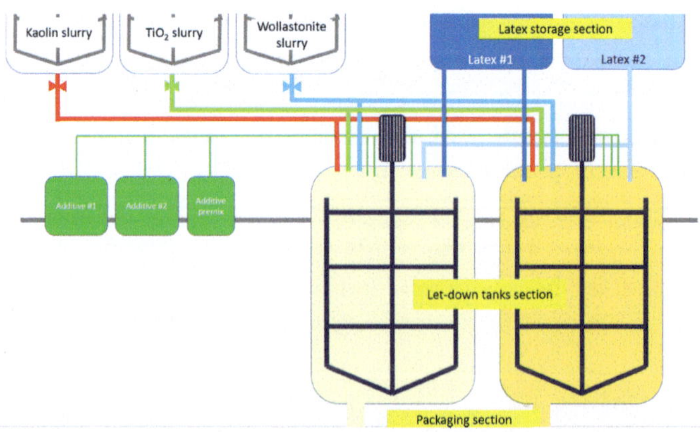

Figure 7.4. Slurry process – stage II – let-down.

Source: Artur Palasz, Creative Commons License.

Furthermore, forming a robust film is not excessively arduous and time-consuming. Adequate molecule weight of the polymers is imperative to confer strength to the film. Styrene-n-butyl acrylate copolymers are frequently employed as binding agents and are designed to possess a suitable glass transition temperature. Styrene exhibits notable hydrophobicity and confers commendable barrier characteristics in a coating. Methyl methacrylate-n-butyl acrylate copolymers are utilized as binders in outdoor settings due to their superior resistance to degradation caused by UV radiation when compared to styrene-n-butyl acrylate copolymers (Butler et al., 2005).

Pigments, such as titanium dioxide, are particulate substances that impart color, disperse light, and confer opacity. The efficient scattering of light by particles of submicron to micron size and with a refractive index different from that of the binder results in the provision of opacity and hiding power. The critical pigment volume concentration (CPVC) is a significant metric that denotes the volume fraction of pigment at which the pigment particles come into contact, leading to a rapid alteration in properties such as gloss and film strength. In most latex paints, the CPVC is present at a concentration of 40 % by volume or greater. Low-cost extenders like calcium carbonate are utilized alongside the pigment to improve opacity. This practice is commonly observed (Henson et al., 1953). The emulsion polymerization process utilized in the preparation of latex results in the presence of water, which confer liquidity and spreadability to the substance, thereby facilitating its uniform distribution. The shift towards water-based paints is primarily motivated by their lower environmental impact than solvent-based paints. Oil-based paints exhibit superior performance owing to the plasticizing effect of the solvent, which facilitates the film formation process. By formulating the latex with care, it is possible to achieve a level of performance in water-based paints that rivals that of solvent-based paints. Typically, paints are formulated with a water content ranging from 60 – 70 % volume (Kostansek, 2007).

7.2.4. Industrial Coatings

Various latex-based coatings, such as automobile, primers, and industrial coatings, have significant commercial value. This is supported by a reference highlighting their importance in the market. Primers are utilized to establish a foundational layer on a given substrate as a preparatory step for the subsequent topcoat application. The primer is required to exhibit robust adhesion to the substrate, effectively level out the surface roughness, and impart a degree of flexibility to the interface between the base material and the top coat of paint. Applying a primer safeguards the topcoat layer from contamination by any staining from the substrate. This can occur due to various factors, such as water seepage via a wooden substrate. The outermost layer, commonly referred to as the topcoat, is accountable for conferring the intended optical characteristics of the coating. Typically, primers are formulated with acrylic copolymers in elevated concentrations to mitigate the occurrence of defects, such as voids, within the coating and to enhance its flexibility. The primary impetus for adopting water-based automotive coatings is the decreasing prevalence of volatile organic compounds (VOCs). These coatings are predominantly composed of polyurethane, polyacryli, polyester, and epoxy dispersions. An automobile coating system generally comprises four distinct layers. Applying an epoxy-amine electro-coat to the steel metal sheet body enhances its corrosion resistance. Subsequently, a polyester primer is administered through spray coating to enhance adherence to the ensuing layer, the basecoat. The basecoat of a car may consist of various materials such as polyacrylic, polyurethane, or polyester, and is responsible for housing the pigments that impart the car's distinctive color. A clear coat is administered onto the surface, amplifying its luster and chemical durability. Chemical crosslinking is employed in curing automotive coatings, rendering them remarkably robust (Winnik & Feng, 1996; Liu et al., 2001) (Figure 7.5).

Tip

When selecting latexes for industrial coating applications, consider factors such as film formation, adhesion, and durability to ensure optimal performance and protection.

Figure 7.5. Latex based waterproofing industrial coating.

Source: Chitra Insultec Private Limited, Creative Commons License.

7.2.5. Paper Coatings

Emulsion polymers serve a dual purpose as coatings in the paper industry. One approach involves enhancing the hydrophobicity of the paper surface to enhance its wet strength. Applying the hydrophobic layer is commonly referred to as surface sizing. Acrylic-based latexes are commonly employed in this particular stage (Lepoutre, 1989). The secondary function of latexes is to serve as a binding agent for the pigments and additives administered onto the paper's surface. Including pigments is a requisite for imparting color to paper and enhancing its optical characteristics, including gloss and opacity. The presence of latex particles aids in the mitigation of surface roughness, thereby facilitating the deposition of ink on a smoother surface. The coating process of the paper is carried out through a high-speed rolling procedure. Emulsion polymers serve as binding agents in the paper and fiberboard sectors (Li et al., 2019).

7.2.6. Textiles and Nonwovens

Textiles are a type of material that is produced by weaving thread or yarn together. On the other hand, nonwovens are composed of shorter fibers that are randomly and loosely bound together, resulting in a porous structure. Crosslinkable latexes are commonly utilized as binding agents to enhance stability, longevity, chemical resilience, and flame retardancy. Acrylic copolymers represent the predominant category of binders. However, alternative copolymers, including styrene-butadiene and ethylene-vinyl acetate (EVA) copolymers, are frequently employed when the cost or performance necessitates their use. The flexibility or rigidity of the textile is regulated by the glass transition temperature associated with the latex binder (Santos et al., 2021) (Figure 7.6).

Figure 7.6. Latex based nonwovens.

Source: Anh Nhi, Creative Commons License.

Household textiles must possess stain-resistant properties and exhibit durability against frequent washing cycles without experiencing significant deterioration. Latexes are binding agents that secure the interlaced fibers and envelop them with a slender safeguarding coating. The latex binder's hydrophobic nature enhances the fabric's water-repellent properties (Ouederni, 2016).

Nonwovens are produced through wet, dry, or spin-laying operations, using either natural or synthetic fibers to create sheet-like structures. The fibers are interconnected through mechanical entanglement, thermal softening of synthetic materials, or chemical binding with latex compounds. Nonwoven materials produce disposable products, such as the lining in diapers, and durable products, such as carpet backing (Karthik & Rathinamoorthy, 2017).

7.2.7. Carpet Backing Binders

The manufacturing of carpeting is a widespread practice on a global scale. The process of producing carpet involves the construction of a pile by stitching turns of yarn onto an initial backing material. Subsequently, the reverse side of the carpeting is treated with a dual-layered back-coating comprising the pre-coat and the scrim-coat. The two strata are commonly composed of carboxylated styrene-butadiene latexes that have been altered with calcium carbonate additives and poly-acrylic thickening agents. The pre-coat is subjected to a foaming process and subsequently administered through direct-coating. Its primary function is as a binding agent that secures the piling in its designated location. Subsequently, the scrim-coat is administered to affix the additional backing material to the rear of the carpet to confer dimensional stability. The polymer latex back-coating serves multiple purposes, such as mitigating flammability, enhancing water resistance, and regulating odors (Urban et al., 2013).

7.2.8. Adhesives

Adhesives find widespread application across various industries where there is a need for prompt and robust bonding of two materials. These industries include but are not limited to automobiles, buildings, furniture, and packing. The rising stringency of environmental regulations has led to a surge in the usage of waterborne adhesives in recent times. Certain adhesives are designed to create temporary bonds that can be reversed, while others are formulated to establish permanent bonds that cannot be undone. Pressure-sensitive adhesives (PSAs) are frequently utilized in tapes and labels, which belong to the reversible class of bonds. PSAs are typically composed of large-molecule acrylic copolymer dispersions that exhibit high viscosity and are maintained at a temperature well above their glass transition temperature. To achieve optimal performance, it may be necessary to modify the formulation of PSAs by incorporating tackifiers, plasticizers, soaking agents, or thickeners. PSAs exhibit adhesion to the surface of a substrate, thereby imparting cohesive strength. A material's adhesive properties are commonly evaluated based on its tackiness, which refers to its ability to adhere rapidly; its peel strength, which is the force required to remove it; and its shear strength, which measures its ability to resist slipping under tangential loads. These parameters are often used to assess the performance of adhesives (Soubam & Gupta, 2021).

7.3. MEDICAL AND HYGIENE

Latex, specifically natural rubber latex, is a prevalent material utilized in various medical products, with gloves and condoms being the most frequently manufactured items (Palosuo et al., 2011).

7.3.1. Medical Gloves

In the healthcare industry, utility gloves safeguard the user and the person being treated from the risk of infection. Household gloves, commonly called gloves, are utilized in domestic settings to safeguard the user from exposure to water, grime, and cleansing agents. Utility gloves are commonly utilized in industrial settings to protect the user and the product from contaminants such as dirt and infections (Figure 7.7).

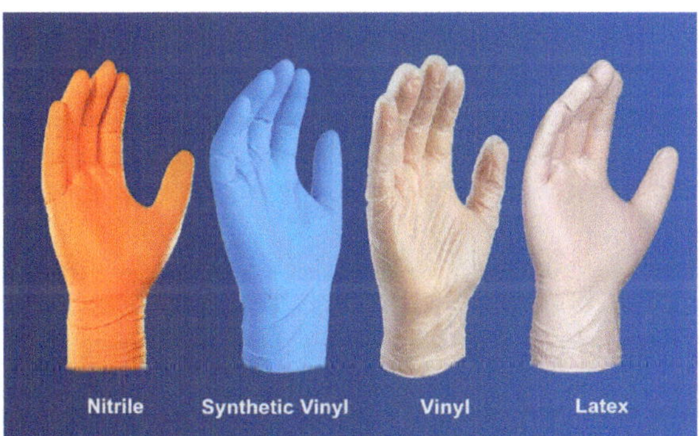

Figure 7.7. Comparison of different types of glovers with latex glovers.

Source: Tom Travin, Creative Commons License.

Professional protective gloves are predominantly manufactured using natural rubber latex (NRL) as the primary material. As previously stated, various utility gloves are produced utilizing NRL. Two distinct market applications can be identified, namely (Mylon et al., 2014):

1. Medical gloves are protective gear commonly used in healthcare settings to prevent the spreading of infections and diseases.
2. Gloves that are not intended for medical use.

Currently, the predominant market application for NRL gloves is in the medical field, specifically for use as medical gloves. The United States of America is the primary consumer of medical gloves. Furthermore, the most active developments have occurred within the domestic market regarding using gloves made by NRL. This report will center on the healthcare glove sector in the United States to demonstrate the potential of NRL gloves (Douglas et al., 1997).

Protective gloves utilized in the healthcare industry are called medical gloves. The various end markets encompass healthcare facilities, healthcare at home, elderly care facilities, emergency vehicles, labs, dentistry and medical offices, emergency services, and other related entities.

Historically, medical gloves were composed of NRL and were coated in fine powder. There exist substitutes for powders in the capacity of glove lubricants. Medical glove lubricants are mandated to perform multiple functions. To begin with, lubricants serve the purpose of acting as mold release agents on glove formers in the process of manufacturing latex gloves. In addition, they are employed as de-tackifying agents to eliminate the inherent tackiness present in latex gloves. Thirdly, glove lubricants aid in donning during surgical and medical operations by offering internal lubrication (Anedda et al., 2020).

The initial substance for applying powder onto natural latex rubber gloves was Lycopodium spores, called club moss or sphagnum moss. Subsequently, it was discovered that this particular lubricant possessed toxic properties, rendering it unsuitable for application for gloves. Consequently, talcum powder, also known as talc, was introduced as a substitute due to its non-absorbable lubricating properties. A substitute of an absorbable lubricant, modified cornstarch, was formulated after clinical complications linked to these agents. During the initial years of the 1970s, a considerable number of manufacturers of medical gloves had substituted talc with a modified version of cornstarch. The present variety of cornstarch powder in use is a crosslinked derivative that has undergone treatment with epichlorohydrin (Korniewicz et al., 2002). The phenomenon of protein transfer from latex to powder is widely acknowledged. During the process of wearing and removing pulverized gloves, the protein-coated particles can become airborne dust, leading to the contamination of large volumes and potential allergic reactions. Numerous states in the United States have put forth proposals to prohibit the utilization of powdered gloves (Crippa et al., 2003).

Currently, gloves manufacturers endeavor to minimize powder use by producing either powder-free gloves or containing reduced amounts of powder. Typically, the powder is eliminated through water washing, further reducing protein content via extraction. Additional techniques for protein elimination in latex processing involve the utilization of specific ion exchange resins, chlorination, or modifying protein composition through protease enzymes (Zare et al., 2021).

The market demand for medical gloves experienced a significant surge during the early 1990s, primarily due to the heightened risk of infectious diseases among healthcare workers, such as AIDS, hepatitis, and the like. Wearing gloves has become a standardized practice across various medical scenarios (Knudsen et al., 2000).

The 1990s witnessed a substantial surge in demand for healthcare gloves, leading to a shortage of gloves due to the supply being unable to keep up with the demand. Consequently, numerous gloves suppliers that do not conform to traditional norms have emerged in the market, particularly from the Asian region. The gloves sourced from non-traditional means exhibited subpar quality in certain instances, particularly regarding the levels of free protein present in the merchandise (Rego & Roley, 1999).

After this period, there was a significant increase in documented hypersensitivity reactions to NRL gloves and other medical equipment that incorporate latex. Research findings suggest that approximately 10% or higher of healthcare professionals in the United States may currently exhibit sensitivity to latex. The issue, as mentioned above, has led to health concerns for workers and patients and occupational complications. Consequently, legal action has been taken, legislation and regulations have been scrutinized by governing bodies, and non-governmental organizations focused on allergy-related matters have been established in various nations, particularly the United States (Carré et al., 2017).

The pressures mentioned above have prompted notable alterations in the composition of materials employed in medical gloves production. The predominant change observed pertains to the transition towards gloves that are devoid of powder. A discernible shift towards using synthetic materials, particularly nitrile, and vinyl, and to a lesser degree, polyurethane, SBR, and polychloroprene, has been observed. Initially, replacing natural materials with synthetic ones was limited due to the substantial price difference between synthetic gloves and latex-based natural rubber gloves in an industry that prioritizes cost. However, the price gap has decreased, leading to a notable shift towards synthetic materials. Additional obstacles encompassed the insufficiency of appropriate grades of synthetic substances for producing medical gloves. Previously, there existed a sole provider of nitrile latex that met the standards for glove manufacturing. Currently, there exist either three or four. The reduction in price disparity, coupled with the increased accessibility of diverse materials and the inclination of user entities to circumvent issues related to NRL hypersensitivity, has expedited the process of glove modification and substitution of materials (Kumar, 2012).

The modifications in glove design and components are undergoing continuous evolution. Significant ambiguity exists regarding the influence of allergy concerns on the utilization of NRL in gloves. The United States and other regions worldwide are engaged in ongoing discussions regarding bans, heightened standards and regulations, litigation, and other restrictions. Producers of synthetic latex are consistently making gradual advancements in their materials and are actively striving to gain a foothold in the market for natural rubber (Dillard).

The medical glove sale exhibits segmentation into various subcategories based on the glove's intended use and structural composition.

Disposable gloves, typically non-sterile, are utilized by diverse healthcare professionals such as nurses, laboratory technicians, and other related personnel for examinations and procedures. The market above segment constitutes the majority, accounting for more than 95% of the medical glove market. The products above are typically characterized by their lightweight construction, with an approximate weight of 8 grams. They are designed for single-use purposes and can be utilized by left- and right-handed individuals. It should be noted that these products are not sterile and may come in either powdered or non-powdered form, with the former facilitating ease of donning.

Surgeons, along with support staff, commonly utilize surgical gloves. This segment constitutes a minor proportion of the overall medical gloves market, accounting for less than 5%. The products above are typically characterized by a heavier gauge of approximately 12 grams, disposability, ambidexterity, sterility, and powder coating to facilitate ease of donning (Jones et al., 1994).

Apart from the aforementioned conventional categories, various specialized types of gloves exist, such as gloves containing lead for radiological purposes, gloves with extended lengths, and so on. The healthcare industry extensively employs standard protective gloves, such as restaurant-grade vinyl gloves, for catering, cleaning, and housekeeping services (Zare et al., 2021).

In medical gloves, a pertinent categorization of materials is the distinction between synthetic and natural gloves. The conventional preference for exam and surgical gloves is natural latex rubber gloves. Artificial gloves are produced using various materials, including but not limited to polyurethane, nitrile rubber, and polyvinyl chloride. Additional materials employed encompass chloroprene, styrene-butadiene rubber, and combinations thereof. Several subdivisions can be identified within these overarching categories, including (Kramer & Assadian, 2016):

Powdered gloves made of NRL and polyvinyl chloride (PVC) contain cornstarch, oatmeal, gypsum, or other substances.

a. The utilization of powder-free gloves is recommended for all categories of materials.

b. One example of lined gloves is those made of latex and lined with polyurethane (PU).

c. Lubricated gloves, such as those containing aloe, are a type of gloves that have been treated with a lubricating substance (Mylon et al., 2016).

All of these variations pertain primarily to the ease of donning. Currently, a diverse array of glove varieties differ in their design, functional characteristics, and cost. Typically, medical gloves lack a woven fabric lining and are considered non-supported. The yearly market for gloves for medical use in North America is approximately 14 billion pairs, with a market worth exceeding $2.5 billion (Yip & Cacioli, 2002).

As previously stated, the market experienced a substantial expansion in the early 1990s. Subsequently, the rate of growth has reached a point of stabilization. The present growth rates are approximated to be between 3–4% annually in volume, while in value, the rates are lower due to the general decrease in unit prices. Brief surges in demand were observed in 2001 and 2000, coinciding with the 9/11 terrorist attacks and the anthrax outbreaks (Pitten et al., 2000).

Approximately 65% of all gloves in the US medical glove market comprise natural latex gloves, signifying their continued dominance in this sector. Replacing NRL with synthetic alternatives has been a noteworthy development, particularly because latex gloves accounted for nearly 90% of all medical gear in the mid-1980s. The proportion of powdered latex gloves has significantly decreased over time. It constitutes merely 30% of the market and exhibits a persistent downward trend. The natural rubber glove value chain exhibits a high degree of complexity (Edlich et al., 2009).

Furthermore, the production of medical gloves has an international scope, with approximately 90% of these products being produced in Asia and subsequently exported to Europe or the USA. Several Asian suppliers are under the ownership of North American glove groups, such as Allegiance, while others are subcontractors or independent Asian producers, such as WRP. The manufacturing of NRL gloves and nitrile gloves is predominantly centralized in the South East Asian region, specifically in countries such as Thailand, Malaysia, China, and Indonesia. A significant proportion of medical gloves utilized in North America are sourced from Asian countries, exceeding 90%. In the present supply and demand cycle stage, the manufacturing sector is deemed relatively unremunerative, prompting significant glove marketers to outsource a growing proportion of their production, particularly to Asian firms. The notable extent of cross-selling and subcontracting activities heightens the complexity of the supply side. Typically, the primary purveyors of medical gloves possess several glove manufacturing facilities but also engage in a significant degree of subcontracting for production purposes (Preece et al., 2021).

The primary providers of NRL are Firestone and Guthrie. There has been significant consolidation within the medical glove industry on the production side. Recent mergers and acquisitions in the glove industry include Ansell's acquisition of the J & J glove business and Kimberly Clark's acquisition of Safe skin (Russell-Fell, 2000).

In recent years, the healthcare industry has undergone significant consolidation and remains the primary medical glove consumer. Hospitals are implementing rationalization strategies in their procurement processes by leveraging group purchasing networks. This can be achieved by establishing a centralized purchasing center that serves multiple facilities or by joining a Group Purchasing Organization (GPO) that negotiates contracts on behalf of its members. Presently, GPOs possess agreements for healthcare establishments that account for more than 80% of the aggregate medical units in the United States of America. The dominance of the leading GPOs in the market and their proportionate share of the overall medical supplies market in the United States is on a steady incline. The amalgamation of GPO groups has expedited the process, resulting in a dominant oligopoly comprising major players such as Premier, Novation, Amerinet, Consorta, and Medassets. The healthcare industry has experienced an escalation in cost pressure in recent times. Considerable endeavors are undertaken to curtail expenses via rationalization, cost containment, and network initiatives. Nevertheless, expenses are continuously increasing due to the influence of customer expectations, legal requirements regarding the quality of services, and legal disputes (Swanson et al., 1994).

The prevailing price range for most NRL medical testing gloves is between $0.06 and $0.11 per glove. In contrast, synthetic nitrile gloves are priced at $0.11 to $0.17 per glove. The cost of a pair of surgical gloves falls within the price range of $1 to $5. In recent years, there has been a downward trend in prices, which is projected to persist (Landeck et al., 2015).

The prices of nitrile gloves have experienced significant reductions, resulting in a gradual convergence of prices between synthetic and natural gloves. The transition in the latex industry from powdered to powder-free products has mitigated the underlying decrease in prices (Mellström & Boman, 2000).

The selection of gloves in the medical industry is influenced by various factors such as cost, efficacy, adverse effects such as allergic responses and dermatitis, and customary practices. Numerous analyzes have been conducted to evaluate the functional efficacy of diverse glove materials. The table the Maryland School of Nursing provided is widely regarded as a thorough and impartial summary (Filon & Radman, 2006).

Each material presents distinct properties, benefits, and drawbacks. Nevertheless, it is possible to draw certain generalizations regarding the practicality of materials. These materials have undergone approval processes and are currently accepted and utilized to a considerable extent. Therefore, none of these options exhibit a critical deficiency that would render them unsuitable for the marketplace (Yunginger et al., 1994).

It is imperative to bear in mind that the primary functional necessity of gloves is to serve as a hindrance against microbial and chemical contamination. Chemical resistance varies among the gloves available in the current market, although all can achieve this. Differences among gloves are observed in their level of comfort, ease of donning, fatigue, and tactility (Taylor & Erkek, 2004).

The NRL serves as the standard for performance in various aspects. The sole areas in which it may be deemed 'substandard' in comparison to its synthetic equivalents are its potential for eliciting allergic reactions and conceivably about certain facets of its tear/puncture resistance (in comparison to PU) and chemical resistance (in comparison to nitrile). Significant advancements have been made in enhancing the efficacy of synthetic materials for glove application over the years. This holds for vinyl compounds as well as for nitrile compounds. The ongoing upgrading process is anticipated to further enhance their competitiveness with NRL (Raulf, 2014).

7.3.2. Condoms

Before the early to the middle 1980s, condoms were considered inconspicuous and underappreciated items. The main purpose for which they were employed centered on preventing conception. The role of condoms in preventing STDs, or sexually transmitted diseases, has been acknowledged since ancient Roman times. However, it was not a significant concern among the general populace. The emergence of AIDS, or acquired immunodeficiency syndrome, and the heightened prevalence of STDs prompted a more candid discourse on sexual topics and the dissemination of productive methods for preventing infection. This facilitated the discussion of condoms beyond the confines of the barber's shop and into the realm of public discourse. In contemporary times, condoms have been reclassified as medical devices. They are subject to stringent regulations governing their production and distribution, rendering them no longer relegated to clandestine commodities (Finger, 1998).

Did You Know?

Latex is commonly used in the manufacturing of condoms due to its excellent barrier properties, flexibility, and biocompatibility.

Numerous nations have implemented domestic regulations about condoms for an extended duration. However, in recent times, these standards have become considerably more rigorous, mirroring the heightened significance of the product. The International Standards Organization (ISO) released a standard for condoms in 1990 after prolonged deliberation in draft form. Subsequently, the European Commission has also adopted the action above, resulting in the approval of the CEN standard (EN 600). Its imminent release is anticipated. As of mid-1998, all condoms marketed within the European Union were mandated to adhere to the specifications outlined in the Commission's Medical Device Directive and exhibit the CE marking. The present global output of marked condoms ranges from 8 billion to 10 billion units annually, with a production potential of 12 billion. However, due to the lightweight nature of a single condom, weighing only 1.5 grams, the overall rubber consumption is a mere 15,000 tons. Apart from Japan and the United States, nations manufacturing a substantial quantity of condoms comprise India, Malaysia, and South Korea. The annual production of condoms for domestic consumption in China amounts to 1.2 billion units. In the 1990s, there was a notable rise in condom manufacturers, particularly in the Asian region (Grimley et al., 1997).

Apart from the market for branded condoms, there exists a substantial market for not branded condoms that are utilized in medical assistance programs, estimated to be approximately 2 billion units. Quantifying the condom market poses a challenge due to donations and substantial government support for producers in China and India, which boast sizeable populations and significant condom production capacities (Fehr et al., 2015).

According to estimates by John Hopkins University, the global demand for condoms is approximately 24 billion units annually. According to reports, the current influx of donations to Africa (estimated to exceed one billion units) has resulted in four free condoms per annum to every sexually active male on the continent. The global market value, exclusive of China, is estimated to be approximately $500 million, with nearly $300 million in the United States. The data above illustrates the impact of governmental subsidies on the market valuation (Rannie & Craig, 1997).

Currently, condoms made from NRL are the primary material used, constituting over 80% of all condoms produced. Condoms are manufactured on dip-molding lines utilizing NRL. The inherent characteristics of natural rubber make it a highly suitable material for producing condoms, exhibiting exceptional tensile strengths exceeding 30 MNm-2 and a point of breaking elongation surpassing 800% (Woolf & Maisto, 2008).

Sensitive individuals may experience skin irritation or allergic reactions when using condoms made from natural rubber. The conditions above result from either the chemical agents utilized in the vulcanization process of rubber or the proteins inherent in the latex of natural rubber. In reality, the issue of allergy has not posed a substantial obstacle in the utilization of condoms, save for a minute subset of individuals who exhibit heightened sensitivity. The frequency of said illnesses has been observed to be notably higher among individuals who use rubber latex gloves, plausibly due to the extended duration of glove usage and the thicker material of the gloves, which necessitates more efficient washing to eliminate any remaining chemicals and proteins (Steiner et al., 2003).

According to market research, a significant reason for the underutilization of condoms is the perception that they diminish sensitivity and sensation during sexual intercourse. The customary range for the wall thickness of latex condoms is between 0.06 mm and 0.07 mm. To achieve a substantial reduction while ensuring safety, it is imperative to employ higher-strength materials. This phenomenon has prompted numerous manufacturers to conduct research on substitute materials throughout the years. In contemporary times, the quest for a solution has gained further momentum due to the emergence of latex protein allergy.

Polyurethane possesses a tensile strength that is twice as high as that of NRL film. This property enables the production of condoms that are considerably thinner than conventional high-quality latex condoms while maintaining comparable durability.

The Duron brand of polyurethane condoms has been found to possess the properties of being odorless, tasteless, and colorless. Biocompatibility testing has indicated that these condoms are non-allergenic. Several organizations and companies are currently engaged in the development of condoms made from alternative synthetic materials. Additional materials that have been evaluated for the production of condoms are styrene block copolymers, including styrene-ethylenebutadiene styrene (SEBS), styrene-butadiene-styrene (SBS), and styrene-isoprene-styrene (SIS). Carter Wallace, Sagami, and SSLI are among the primary manufacturers of condoms (Callahan et al., 2000).

7.3.3. Personal Care and Hygiene Products

Personal grooming and hygiene products are essential for maintaining cleanliness, comfort, and well-being. The utilization of latex materials in this industry has exhibited a wide range of applications, providing notable characteristics such as resilience, longevity, and imperviousness. This section delves into the extended application of latex in personal care and hygiene commodities, encompassing latex air balloons, rubber bands and buckles, cloth diapers, women's hygiene products, and items for incontinence (Allmers et al., 2002).

7.3.3.1. Latex Balloon

Rubber balloons are a frequently selected option for social gatherings, festivities, and ornamental intentions. The physical properties of latex, specifically its elasticity and stretchability, render it a highly suitable substance for fabricating balloons that can be effortlessly inflated and shaped into diverse configurations. Latex balloons possess the characteristics of being lightweight, displaying vibrant colors, and exhibiting exceptional sealing properties. These items' extensive utilization in formal occasions and casual festivities can be attributed to their adaptability, cost-effectiveness, and eco-friendliness (Baldi et al., 2003) (Figure 7.8).

Figure 7.8. Multicolor latex balloons.

Source: Oriental Trading, Creative Commons License.

7.3.3.2. Elastic Bands and Straps

Flexible bands and straps made of latex are frequently employed in personal care merchandise, including but not limited to hair accessories, clothing, and medical equipment. The innate elasticity of latex enables the bands above and buckles to furnish a snug and agreeable fit. These items find application in various wearable accessories such as headbands, hair ties, waistbands, and bra straps, providing a secure and pliable grip. Latex elastic bands are utilized in medical contexts to provide gentle and effective pressure and support in various applications such as wrappings, skeletal supports, and compression garments (Mason & Rosa, 2022).

7.3.3.3. Disposable Diapers

Disposable diapers' effectiveness in maintaining infants' dryness and comfort is attributed to the absorbent and liquid-repelling characteristics of latex. Latex is an elastic material for the leg handcuffs and waist bands of diapers, promoting a snug fit and impeding leakage. The elastic properties of latex facilitate a close-fitting contour around the infant's physique, ensuring comfort and impeding the leakage of bodily excretions. Latex contributes to preserving the form and structural stability of the diaper throughout its usage (Counts et al., 2017).

7.3.3.4. Feminine Hygiene Products

Latex is a prevalent material utilized in producing feminine hygiene items, including feminine hygiene products and panty liners. Latex is utilized to configure elastic bands and glue constituents to guarantee a steadfast fit and avert seepage. The flexibility and suppleness of latex offer a comfortable and adaptable experience, enabling these items to adapt to the anatomical structure of the body and furnish dependable safeguarding (Spaner et al., 1989).

7.3.3.5. Incontinence Products

Latex materials are utilized to manufacture incontinence products, such as protective underwear, adult diapers, and bed pads. The products in question utilize latex elastic straps and waistbands to ensure a secure fit and mitigate the risk of leakage. The stretchability and resilience of Latex aid in preserving the product's structural soundness, even when subjected to motion, thereby guaranteeing comfort and imperviousness for those with incontinence requirements (Cottenden, 1988).

The increased utilization of latex in hygiene and personal care commodities exemplifies its adaptability and appropriateness for functions necessitating flexibility, resilience, and obstructive characteristics. Latex materials are widely used in various personal care items such as air balloons, disposable elastic bands, disposable diapers, feminine hygiene products, and incontinence products. These materials enhance these essential items' performance, comfort, and overall efficacy. Notably, certain individuals may exhibit hypersensitivity to latex, hence necessitating the provision of substitute substances or latex-free options to cater to their requirements. Manufacturers and brands consistently endeavored to create substitutes free of latex yet still possess the desired characteristics and functionalities of hygiene and personal care products. This is done to accommodate a broader spectrum of users (Dingwall, 2008).

7.3.4. Other Medical and Hygiene Applications

The latex market is significant in producing disposable anesthesia breathing bags. The natural latex market currently accounts for 85% of the total market share. However, the market share of synthetic latex is anticipated to witness a substantial surge in the forthcoming years (Singer & Plotz, 1956).

Various medical applications include balloons of different sizes, cuffs, covering, specimen bags, probe encompasses, diaphragms, catheter air balloons, catheter sleeves, and contraception diaphragms. The utilization of natural rubber tubing made of latex has been widespread in the healthcare sector for a considerable period (Grajales et al., 2014). Subsequently, it has increased into various industries such as dentistry, food, toys, instrumentation, and sporting goods. Latex is widely recognized for its exceptional flexibility, rendering it one of the most pliable and resilient rubber materials. Latex tubing exhibits resiliency, enabling it to retain its memory even after being subjected to multiple stretching, bending, or pulsating forces. Latex exhibits high resistance to multiple sterilization methods, including steam, ethylene oxide, and gamma radiation. The absence of plasticizers, coagulants, or fillers prevents the migration and contamination of fluid streams and the hardening of tubes (Husain et al., 2008).

7.4. SPORTS AND RECREATIONAL EQUIPMENT

Using specialized equipment is frequently necessary for recreational and athletic endeavors to augment performance, guarantee safety, and optimize functionality. The utilization of latex materials has been observed in diverse sports and amateur. The equipment provides distinctive characteristics that enhance performance, longevity, and user satisfaction. This section delves into latex application in various sports equipment such as swim caps, physical activity bands, obstruction tubes, and the coatings over balls and rackets (Laad, 2020).

7.4.1. Latex Swim Caps

Latex swim caps are prevalent in aquatic activities such as swimming and other water sports. Swimming caps offer many advantages to swimmers, such as diminished water resistance, heightened hydrodynamic efficiency, and enhanced safeguarding of hair. The physical properties of latex, such as its elasticity and ability to conform to the body, facilitate a close and stable fit, thereby reducing hydrodynamic drag and enhancing swimmers' aquatic mobility. Latex swim caps are characterized by their lightweight nature, comfortability during use, and ability to offer safeguards against chlorine and UV exposure (Sebestyen, 1997) (Figure 7.9).

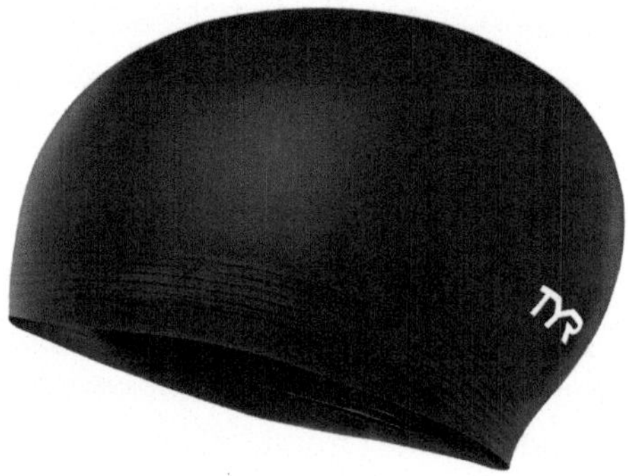

Figure 7.9. Schematic of the latex swim caps.

Source: TYR, Creative Commons License.

7.4.2. Latex Exercise Bands

Latex exercise bands, also referred to as bands of resistance or exercise bands, are multifaceted and efficacious instruments employed to enhance muscular strength, aid in rehabilitation, and facilitate physical conditioning. The bands provide diverse degrees of resistance, enabling individuals to focus on distinct muscle groups and personalize their exercise routines. Latex bands are a favored option due to their inherent elasticity, durability, and capacity to stretch without experiencing a reduction in tension over an extended period. Resistance training equipment offers a secure and practical alternative for individuals seeking to engage in diverse exercises, both in the comfort of their own homes and within fitness establishments (Kusumawati et al., 2022) (Figure 7.10).

Figure 7.10. Image of latex exercise bands.

Source: Ambitross Private Limited, Creative Commons License.

7.4.3. Latex Resistance Tubes

Latex obstruction tubes are pliable and resilient instruments for resistance exercises and strength training, akin to exercise bands. The aforementioned cylindrical structures, frequently furnished with handles or supplementary components, furnish oppositional force in numerous orientations and facilitate diverse physical activities that concentrate on distinct muscular clusters. Utilizing latex resistance tubes allows users to modify resistance levels, thereby facilitating progression in workouts and stimulating the development of strength and endurance. The elasticity and strength of latex ensure compatible tension and dependable performance during workouts (Kluemper et al., 2006).

7.4.4. Latex Coatings for Balls and Rackets

Latex coatings are frequently utilized in sports equipment, including balls and rackets, for various sports such as volleyball, squash, basketball, and racquetball. Utilizing latex coatings yields superior traction, manipulation, and rotation, augmenting the athletic proficiency of players during gameplay.

Latex coatings provide an adhesive surface that enhances ball handling, enabling athletes to produce increased rotation and superior agility. Latex coatings applied to racket handles in racket sports have been found to improve grip, thereby minimizing slippage during high-intensity gameplay. This, in turn, enables players to sustain precision and control in their shots.

The versatility and value of latex in improving efficiency and user experience are exemplified by its application in sports and recreational equipment. Latex materials possess distinctive characteristics such as resilience, robustness, and adhesion, ensuring maximum efficiency and protection in diverse athletic and leisure pursuits, including swim caps, physical activity bands, resistance tubes, and ball and racket coatings. Notably, although latex materials offer considerable advantages, certain individuals may exhibit hypersensitivity reactions to latex. Hence, it is imperative to contemplate substitute materials or furnish alternatives devoid of latex for individuals with hypersensitivities. Manufacturers and sports equipment providers consistently innovate to create safer and more comprehensive alternatives that meet the needs of diverse individuals and their preferences (Jafarnezhadgero et al., 2022).

7.5. BUILDING AND CONSTRUCTION

7.5.1. Concrete Modification

The addition of polymer can be utilized to modify the properties of concrete. There exist two distinct categories of altered goods:

Polymer-impregnated concrete is a type of concrete that has been treated with a polymer material.

Polymer cement concrete is a type of concrete that incorporates polymers, which are long chains of repeating units, into the cement mixture.

The initial step involves saturating precast, solidified concrete made from Portland cement with a monomer, which is subsequently transformed into a solid polymer, resulting in the first type. The second type of concrete is generated by substituting a portion of the cement binders in the concrete mixture with polymer in the form of latex. Both materials exhibit superior strength, reduced water permeability, heightened chemical resistance, and increased freeze-thaw durability compared to traditional concrete. Impregnated concrete is typically associated with high costs and is primarily utilized for specialized purposes.

The incorporation of polymer latex into concrete leads to significant enhancement of its properties while maintaining cost-effectiveness. A diverse array of latices are employed to produce polymer cement concrete items and mortars among the frequently utilized latices, Polyvinyl acetate, Acrylics, vinyl chloride copolymers, SBR, nitrile, polyvinylidene chloride, and natural rubber. Individual polymers generate distinct physical properties.

Using acrylic latex fosters a highly effective water-resistant linkage between the altering polymer and the concrete's constituents. Conversely, applying latices composed of styrene-based polymers yields substantial compressive strength. The curing process of latex polymer concrete made from cement differs from that of traditional concrete due to the formation of a polymer film on the product's surface. This film retains a portion of the moisture required for uninterrupted cement hydration. The latex product's film-forming property typically results in a shorter moist curing duration than traditional concrete. Polymer cement concrete that incorporates polymer latex typically demonstrates superior adhesion to steel support and pre-existing concrete, favorable ductility, resilience against water and water-based salt solutions, and immunity to harm from freeze-thaw cycles. Modified concrete's flexural durability and strength are typically superior to those of unaltered concrete. The elastic modulus of concrete may exhibit variability in either direction relative to unmodified concrete, contingent upon the specific polymer latex employed. As an illustration, a decrease in modulus is observed with an increase in the elasticity of the polymer. In general, the incorporation of polymer into polymer cement concrete results in forming a low modulus phase, which is associated with higher levels of creep compared to plain concrete. The creep reduction magnitude depends on the type of polymer latex used, with a decreasing order of effectiveness observed. Regarding performance ranking, the order of materials is as follows: unmodified cement, polyvinylidene chloride, SBR, and acrylic (Warson & Finch, 2001).

The primary utilization of polymer cement concrete that contains latex is in the domain of floor surfacing, owing to its cost-effectiveness and non-dispersive nature. Due to its reduced shrinkage, notable resistance to permeation from diverse liquids like water and salt solutions, and commendable adhesive characteristics to pre-existing concrete, this material is especially fitting for slender (up to 25 mm) floor toppings, overlays for concrete bridge decks, anti-corrosive overlays, concrete repairs, and patching. Utilizing dispersible powders that can form latices when mixed with water is a comparatively modern advancement. Redispersible polymer powders are produced through the process of spray-drying aqueous dispersions. The dispersions commonly utilized are founded on homopolymers of vinyl acetate, copolymers of vinyl acetate and ethylene, as well as styrene-acrylate and SBR. To enhance the effectiveness of the spray drying process and improve the dispersibility of the final product, water-soluble protective colloids are added to the dispersion post-production. Anti-caking agents are incorporated into the spray drying procedure to prevent the coalescence of the ultimate discrete powder entities.

Like other constituents of mortar, they can amalgamate with sturdy construction materials, conferring a substantial technological benefit for their utilization. Redispersible polymer powders are primarily utilized in dry blend products, including thermally insulated systems, cement-based tile adhesives, self-leveling carpeting compounds, grout modification, paints, and plasters.

Contemporary thin-set cement mortars and grouts composed of polymer latex comprise an acrylic additive in the form of a desiccated powder, which is incorporated with the other dry constituents during manufacturing. Polymer-modified thinsets and grouts, when transported in dry form to the work location, incur lower shipping costs than liquid additives predominantly composed of water. The activation of polymer-modified mortars solely requires water. The visual determination of the concentration of liquid additives poses a challenge. It is not easy to discern whether the additive is at full potency or has undergone dilution with water (Awoyera et al., 2023).

7.5.2. Asphalt Modification

Asphalt mixtures are composed of a binding material, namely asphalt, which is blended with aggregate. Asphalt road surfaces tend to soften under intense solar radiation during hot summer days, while conversely, they become exceedingly brittle during cold winter nights. The phenomenon of 'rutting' results from the permanent deformation of the soft asphalt pavement due to heavy traffic. Pavement cracking occurs in the winter due to the loss of flexibility in the asphalt binder.

The statement above implies that the efficacy of the asphalt binder is contingent upon its viscoelasticity, which must be maintained within a specific application timeframe to dissipate the stress induced by vehicular traffic effectively. The application window can be extended through polymer modification, primarily by enhancing viscoelasticity under high-temperature conditions. The application of modified asphalt has been observed to enhance the fatigue resistance of pavements, thereby increasing their lifespan. For instance, modified asphalt has been shown to extend the lifespan of pavements to 15 years, as opposed to 10 years without modification.

Two primary methodologies exist for fabricating asphalt pavement, namely 'Hot Mix' and 'Cold Paving.' In the hot mix procedure, the aggregates undergo heating to a temperature exceeding 200°C to eliminate any remaining moisture and are subsequently blended with molten asphalt at a temperature greater than 165°C.

The predominant source of energy consumption in the hot mix process is attributed to the heating of the aggregate, which constitutes approximately 90% of the total energy usage. The primary polymer modifier utilized in this system is the styrene-butadiene-styrene terpolymer, commonly recognized as Kraton®. The commonly employed 'chip-seal' technique utilizes a cold paving method involving an asphalt emulsion of 65–75% dispersed in water. The utilization of cold aggregates has the potential to decrease energy consumption substantially. Research has demonstrated that SBR latices are effective hot and cold procedure modifiers (Wen et al., 2017).

High pH SBR emulsions stabilized with anionic surfactants are utilized for hot mix applications. The customary proportion of butadiene is 76%, accompanied by a solid content ranging from 65% to 70%. Applying heat and abrasion over time results in the formation of a uniform mixture of polymer within the asphalt. Upon the addition of latex, the polymer particles undergo swelling due to the heat and solvating impacts of the polymers and oils present in the asphalt. The process of cold paving involves the utilization of SBR emulsions that have been stabilized with cationic surfactants, which are similar. The SBR utilized for asphalt modification is engineered to produce a polymer film devoid of coagulum formation, thereby facilitating the acceleration of early strength development. The migration of latex particles in the presence of water leads to their accumulation in the menisci, where they undergo a spontaneous transformation into a cohesive polymer film that envelops the asphalt particles. This process can be likened to spot welding, as it effectively enhances the binding power of the asphalt particles. The asphalt emulsion mustn't undergo coalescence to achieve optimal honeycomb structure formation.

SBR emulsions provide several advantages to asphalt.

1. The ability to withstand permanent deformation at high temperatures is crucial to asphalt concrete pavement performance. While it is acknowledged that the binder is not the sole contributor

to rutting and shoving, it is a significant factor. The utilization of SBR latices has been observed to mitigate the occurrence and intensity of permanent deformation by enhancing the binder's elasticity at elevated temperatures. The enhanced flexibility of the binder facilitates a more efficient recovery from deformations compared to unaltered systems.

2. Incorporating styrene-butadiene rubber (SBR) in asphalt modification yields a notable advantage in its ability to withstand low-temperature cracking. This is attributed to the decrease in the improved asphalt's stiffness at low temperatures, consequently mitigating the likelihood of thermal cracking.
3. Incorporating Styrene-Butadiene Rubber (SBR) mitigates the effects of thermal and oxidation aging consequences in the asphalt binder, enhancing its resistance to short- and long-term aging. The utilization of modified asphalts has exhibited a significant improvement in both short and long-term aging, thereby highlighting the enhanced durability of the binder. The mitigation of aging phenomena significantly enhances the pavement's longevity, thereby diminishing the life cycle expenses of the pavement configuration.
4. SBR latices exhibit the ability to mitigate the occurrence of stripping and detachment of the asphalt binder from problematic aggregates. SBR latices are commonly employed by numerous organizations as ingredients in hot-mix asphalt surfaces utilized for both road and airport applications (Azahar et al., 2016).

BASF AG has been employing comprehensive evaluations of manufacturing and disposal procedures to capitalize on products with the highest environmental sustainability and cost-effectiveness. BASF has determined that utilizing cold mix asphalt containing 0.3% SB latex and small emulsifiers is a more economically viable and ecologically sustainable alternative to conventional hot mix asphalt.

The utilization of cold mix presents various benefits, including the absence of emissions, enhanced safety for workers compared to hot mix, and significant energy savings due to its low-temperature application. Furthermore, the utilization of cold mix for road paving results in a considerable reduction in raw material consumption due to the lower volume required compared to hot mix (Yildirim, 2007).

7.5.3. Adhesives and Sealants

Latex polymers are extensively used to formulate adhesives for various applications such as flooring, panel/stud, and contact adhesives by contractors. Polyvinyl acetate is a frequently utilized material in various adhesive applications. However, it may be combined with acrylics when water resistance is necessary. Pure acrylic latices are commonly utilized in caulking and sealant applications.

Remember

Latex-based adhesives and sealants offer advantages such as good adhesion, flexibility, and water resistance, making them suitable for various building and construction applications.

7.6. TEXTILES AND NON-WOVEN FABRICS

Latices play a crucial role in producing both textile and non-woven fabric materials.

7.6.1. Textiles

Conventional textiles are fabricated by interlacing natural or synthetic fibers into yarns, subsequently subjected to a finishing process to yield the final product.

Polymer latices are extensively utilized in finishing operations, serving as:

1. Coatings and back coatings.
2. Pottery or ceramic objects
3. Binders for Pigments
4. Adhesives.

Applying latex coatings to fabric has enhanced its tensile strength, stiffness, and resistance to abrasion. Applying latex-based coatings to the reverse side of the material results in improved physical properties such as increased strength, enhanced durability, and heightened resistance to wear. The tactile perception of fabric, commonly called its 'handle' or 'hand,' encompasses a spectrum of characteristics ranging from pliable and velvety to rigid and unyielding. The selection of latex material significantly impacts the tactile properties of the end product, and the desired hand for a specific application can be achieved by carefully considering the type of latex utilized (Li et al., 2020) (Figure 7.11).

Figure 7.11. Illustration of the lace & latex.

Source: Material District, Creative Commons License.

7.6.2. Non-Woven Fabrics

Non-woven textiles, as their nomenclature suggests, are not produced through the traditional weaving process of textiles. Instead, they are composed of a web or uninterrupted sheet of fibers that are mechanically deposited. The fibers can exhibit a stochastic distribution or a unidirectional alignment upon deposition. The fibers most commonly utilized in various applications comprise cellulosic, polyamide, polyester, polypropylene, and polyethylene. Formulating down spun fibers onto a belt can be achieved through various methods such as carding, air-laying, or wet-laying. The fibers are capable of being drawn during this process.

Before the emergence of contemporary methods like thermal attachment and hydroentangled fabrics, chemical binders were necessary for most non-woven fabrics to confer any degree of structural soundness. This trend persisted for more than four decades. Furthermore, the binder was utilized to impart and communicate various characteristics essential for the textile's efficient functioning. During this timeframe, binders were considered the limiting factor in producing non-woven fabrics that met the required standards. The non-woven industry utilized the same fibers as other industries that relied on textiles and fibers, indicating that these fibers were deemed suitable for use. The binder commonly restricted the non-woven fabric's performance. The shortcomings of non-woven materials were primarily due to an insufficient binding agent. Frequent grievances included:

1. The tensile strength of the fabric is insufficient.
2. The textile material exhibits excessive rigidity.

3. The textile material exhibits an insufficient capacity for absorption.
4. The launderability of the fabric is deemed to be inadequate.
5. The dry cleanability of the fabric is insufficient.
6. The tactile sensation of the fabric does not exhibit the characteristics typically associated with textiles (Sakthivel & Ramachandran, 2012).

Consequently, substantial research has been conducted to enhance the ongoing development of chemical binders. The enhancements in non-woven performance over an extended period were largely attributed to the advancements in the functionality and usefulness of the binder.

During the initial phases of non-woven production, various kinds of natural adhesives and resins were employed to bond non-woven fabrics. Although these webs exhibited certain qualities of integrity and strength, they also possessed several notable shortcomings. As a result, synthetic binders were created to fulfill non-woven textiles' structural and performance prerequisites.

The initial synthetic binder used in significant quantities was vinyl acetate (VA) latex. The substance exhibited markedly enhanced adhesive characteristics, potency, and efficacy compared to the initial naturally occurring adhesives. The binder exhibited notable versatility in its composition, rendering it amenable to diverse application techniques, such as print bonding, for affixing to fiber webs.

The non-woven industry encountered an unavoidable trade-off in the physical characteristics of fabrics bonded with artificial fibers. To enhance the tensile properties of the textile, it is imperative to administer escalating quantities of resin, thereby leading to a concomitant rise in rigidity. The attainment of softness may require the relinquishment of strength.

The advancement of polymer technology in producing artificial binder systems has resulted in an expanded range of chemical components that offer enhanced flexibility concerning binder durability, strength, and other related properties.

Implementing cross-linkable and self-crosslinking binder polymers has resulted in a novel spectrum of fabric characteristics. This was especially significant in durable non-woven materials, where attributes such as washability and dry cleanability held considerable importance (Purdy, 1983).

7.6.3. Important Characteristics of Latices for Textile and Non-Woven Applications

Adhesive binders in textiles and non-woven sectors constitute 5% to 60% of the fabric's dry weight. Hence, they are not solely adhesive agents but impart several crucial attributes to the textile.

1. The influence of the binder may vary depending on the intended application of the fabric.
2. The cohesiveness of the applied binder is closely linked to the strength of a fabric.
3. The properties of firmness and stiffness are of interest.
4. Adhesion is a highly desirable property characterized by a strong adhesive strength to fibers.
5. The attribute of flexibility in textile fibers is frequently sought after, particularly when a supple texture is desired.
6. The quality of a soft hand in textiles can be achieved by incorporating some degree of fiber movement.
7. To prevent irreversible deformation of fabric, it is imperative to possess adequate elastic recovery under strain.
8. The durability of non-woven products during cleaning processes is a crucial aspect that must be considered in their intended applications. Latex binders, especially cross-linked ones, can offer resistance to washing, drying, and cleaning, thereby enhancing the durability of such products.
9. The binder must resist aging, maintaining stability and preventing degradation during storage and utilization within the fabric.
10. The requirement for a varied spectrum of colors and the need for colorfastness and avoidance of yellowing issues are important factors to consider in achieving good color and color retention.
11. The phenomenon of surface tension.
12. Stability of dimensions.
13. Compatibility.
14. The hydrophilic and hydrophobic characteristics of the binder are significantly affected by the type of latex employed (Stuart et al., 2022).

7.6.4. Types of Latex Binders

The ensuing evaluation of various chemical types of latex binders offers insight into their respective efficacy and their strengths and limitations in the context of textile and non-woven uses. The present text alludes to the underlying polymer's glass transition temperature (Tg), which denotes the temperature at which the polymer transitions from a stiff solid to a pliable deformable polymer. This temperature is a measure of the comparative flexibility of a bonded fabric. VA is a chemical compound commonly used in industrial applications. Vinyl acetate binders exhibit a high firmness, with a glass transition temperature (Tg) ranging from +30 °C to +40 °C. These binders are widely utilized due to their relatively low cost. The material exhibits commendable dry strength and resilience, albeit displaying a certain degree of hydrophilicity and proclivity towards yellowing under thermal exposure (Yabuuchi et al., 2015).

7.6.4.1. Acrylic

These binders exhibit superior durability, color stability, and performance in dry and wet conditions. Acrylic binders exhibit the most extensive array of fabric hand characteristics. The formulations can be adjusted to exhibit a wide range of hardness, from low values with a glass transition temperature (Tg) of -40°C to high values with a Tg of 105°C. Utilizing these binders is feasible in almost all non-woven applications, albeit their cost tends to be relatively high. Crosslinking of these polymers can be achieved, resulting in a significant enhancement in their durability.

7.6.4.2. Styrenated Acrylics

These binders exhibit high hydrophobicity and possess a robust nature. The textile's tactile properties exhibit a range of softness to firmness, with its glass transition temperature (Tg) varying between -20°C to +105°C. The binders above are suitable for deployment in scenarios requiring a certain degree of moisture resistance without cross-linking. Utilizing this latex binder entails certain compromises regarding its resistance to UV and solvents 9 Kristoffersson et al., 1998).

7.6.4.3. Vinyl Acrylics

The hydrophobicity of these binders is higher than the straight-up VA binders. These materials exhibit superior properties in terms of durability, flexibility, and enhanced color retention. Hybrid materials represent a viable middle ground between VA and acrylic, exhibiting comparable cost-effectiveness and performance capabilities. The range of hand softness is restricted to intermediate levels of softness (Tg=-10 0c) up to a firm level of softness (Tg=+30 0c).

7.6.4.4. Vinyl Acetate-Ethylene (VAE)

The latex binders exhibit a Tg range spanning from -20°C to +115°C, indicating a soft range comparable to a medium textile hand. These materials demonstrate a notable degree of wet strength and display exceptional absorbent properties. Typically, they exhibit a lower cost compared to acrylics. Compared to other binders, they exhibit a greater propensity for emitting odors. These are primarily used in wipes, air-laid pulp fabrics, and similar products (Lee, 1997).

7.6.4.5. Styrene-Butadiene (SBR)

The binders exhibit a commendable balance between flexibility and durability. The materials exhibit a range of hardness levels, spanning from low values of Tg at -30 DC to high values at +80 DC. The SIB ratio is a prevalent approach for characterizing the relative handle that arises from applying styrene and butadiene binders. An increase in the concentration of styrene results in a corresponding increase in the material's stiffness. Upon crosslinking, this particular binder category exhibits high hydrophobicity and durability. The tendency of oxidation affects them to some extent in the presence of heat and light (Isozumi et al., 2020).

7.6.4.6. Polyvinyl Chloride (PVC)

Polyvinyl chloride's homopolymer is a polymer that is extremely rigid and hard, with a glass transition temperature of +80 DC. To confer flexibility and film-forming characteristics, it is necessary to plasticize this polymer. Typically, the PVC binders employed in non-woven materials are rendered pliable through internal softening via co-polymerization of vinyl chloride with acrylic monomers possessing softer characteristics. Most of these polymers exhibit a relatively rigid hand range, characterized by a glass transition temperature exceeding +30 degrees Celsius. Due to its thermoplastic nature, this polymer exhibits excellent performance in heat and dielectric sealing applications. This attribute can confer a benefit in certain applications.

Furthermore, chlorine within the polymer enhances its ability to resist combustion. The characteristic above constitutes a fundamental advantage of employing this binder category. Nonetheless, the presence of chlorine in the polymer results in a propensity towards yellowing during heat aging, which can be attributed to the expulsion of hydrogen chloride.

7.6.4.7. Ethylene/vinyl Chloride (EVCL)

The binders in this course exhibit a wider range of manual flexibility, spanning from Tg = oDC to +30 DC, and do not necessitate external plasticization agents as observed in PVC binders. The inclusion of chlorine once again imparts a degree of flame resistance. The binders demonstrate commendable acid resistance, moderate water resistance, and exceptional affinity towards synthetic fibers. There is a propensity for yellowing as an object undergoes the process of aging. The material in question can be described as an internally PVC binder, with the ethylene monomer as the softening agent (Najafi et al., 2018).

7.6.5. Manufacturing of Non-Wovens

The bonding process is typically essential to the non-woven-in-nature web manufacturing process. Certain fabric constructions may employ multiple bonding processes to augment their physical or chemical characteristics. The prevalent bonding techniques comprise saturation, foam, spray, and print bonding.

7.6.5.1. Saturation Bonding

The process of saturation bonding involves complete immersion of a web in an adhesive bath or flooding of the web at the nip point of a pressure roll set. The removal of excess binder is achieved through the application of vacuum or roll stress.

Drying and curing can be conducted through steam-heated drying cans, air ovens, or perforate drum dryers. The levels of binder addition span from 20% to 60%. The method under consideration offers several benefits, including ease of use, the ability to regulate tensile strength, and the option to adjust softness through the selection and quantity of binders. The drawbacks entail the significant impact of binders on the material's flexibility and the constraints on its loftiness, encompassing bulk and resilience (Butler et al., 2005).

7.6.5.2. Foam Bonding

Foam bonding is used to administer binder at reduced water and elevated binder-solids concentration levels. The fundamental principle utilized entails the utilization of air and water as the binding agent solvent and conveyance medium. The utilization of foam bonding in non-woven materials results in a reduction of energy consumption during the drying process, as a smaller amount of water is utilized. Foam production is achieved by introducing air into the developed latex, coupled with mechanical agitation of the binder solution. The foam can withstand collapse throughout the application and curing process by incorporating a stabilizing agent into the binder solution.

Consequently, the resulting bonded fabric will demonstrate improved loft, hand, and resilience. The benefits encompass reduced energy consumption for web drying, mitigated binder migration, and regulated softness through the selection and quantity of binders. The drawbacks of this approach pertain to challenges in regulating the procedure and achieving sufficient levels of foaming (Bonsignore, 1979).

7.6.5.3. Spray Bonding

Spray bonding involves the application of binders onto webs in motion through the use of spraying techniques. The process of spray bonding finds its utility in fabric-based applications that necessitate the sustenance of elevated loft levels. The atomization process is utilized to apply the binder onto the topmost layer of the web in the form of fine droplets. This can be achieved through air pressure, hydraulic pressure, or centrifugal force and is facilitated by a system of nozzles. Adding a binder to the lower surface of a web can be achieved by reversing the direction of a second conveyance and traveling the web underneath a second spray station. Following each spraying cycle, the web undergoes a thermal treatment process to eliminate moisture, and subsequently, the binding agent is subjected to a tertiary thermal treatment process for curing (Kristoffersson & Carlström, 1997).

7.6.5.4. Print Bonding

Print bonding involves the selective application of binder material to predetermined areas of the fabric. This technique is commonly employed in fabric applications that necessitate the absence of binder material in certain regions, such as cover stocks and wipes. A significant number of lightweight non-woven materials are subjected to print bonding. Printing patterns augment durability, fluid transportation, texture, tactile sensation, absorption, and draping. The process of print bonding is predominantly executed through the utilization of gravure rolls. The levels of binder addition are contingent upon the area and depth of engraving and the level of binder solids. The utilization of rotary screen rolls can lead to enhanced pattern versatility. Drying and curing are typically executed through heated containers or steam-heated cans. Two distinct categories of printers exist rotary display and rotogravure printing devices. In the context of rotary screen printing, binders are administered via a hollow applicator roll, whereas in rotogravure printing, they are dispensed through an imprinted applicator roll. The primary benefit lies in attaining exceptional softness in non-woven textiles while maintaining sufficient tensile strength (Oh et al., 2019).

7.6.6. Other Commercialized Uses

In addition to the aforementioned primary applications, emulsion polymers find utility in various other sectors, which will be briefly alluded to by illustration. Polymer dispersion-based floor polishes are commonly employed to augment the luster and resilience of floor coverings. Printing inks based on Latex are occasionally employed for printing on plastic films, including bags, or paper products, such as cardboard containers. Emulsions find widespread usage in the cosmetics sector. Latex foams with a cellular structure can be generated through frothing, which involves the incorporation of air voids into latexes. This is followed by crosslinking, which stabilizes the foam's structure. Adding latexes to concrete has been observed to enhance its adherence, endurance, and chemical resistance, as noted in previous research. The utilization of polymer dispersions as binders have been found to improve the efficacy of leather finishing procedures. Plastisols, composed of PVC latexes that have undergone spray drying, grinding, and redispersion into organic media, are frequently employed in coatings applications. The addition of latexes to asphalt serves as a means of enhancing its strength and durability (Järnström et al., 2010).

7.7. SPECIALIZED AND POTENTIAL APPLICATIONS

Latexes are being identified as having potential uses in scientific research and specialized areas, such as technological advancement and optoelectronics. The latexes utilized in such applications are commonly homogenous in magnitude, possess a regulated diameter, and may exhibit a particular surface functionalization, such as amino. These latex types are considered specialty products rather than common goods and are often associated with high costs. The distinctive characteristics inherent in these latexes and the adaptable nature of such systems facilitate numerous pragmatic applications. The potential applications mentioned are situated in a transitional phase between conceptualization and full-fledged production. While some are currently undergoing preliminary development, others are being produced on a smaller scale (Rajkuberan et al.,2015).

7.7.1. Latexes as Scientific Tools

Latex particles exhibit exceptional adsorption properties when utilized as surfaces in scientific investigations. Latex particles can adsorb active surface substances, rendering them valuable in quantitative adsorption investigations. Before usage, the particle surfaces must be devoid of adsorbed substances, thereby necessitating their cleansing. Latex particles are valuable model materials for investigating flocculation and film-making phenomena. Introducing functional groups through surface treatments is a feasible approach to modifying latex systems, particularly for applications that are adsorbed by diverse organic and inorganic substances. The outermost layer of latex particles can be covalently attached to these materials by employing innovative surface physics and chemistry techniques (Figure 7.12).

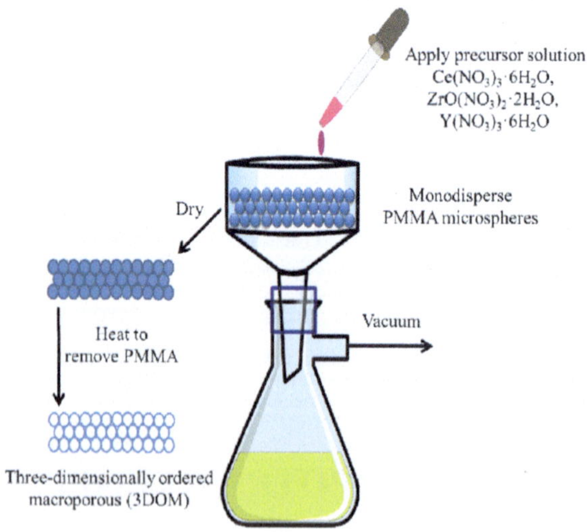

Figure 7.12. The schematic equipment used for the synthesis of monodisperse latexes.

Source: Hongxing Dai, Creative Commons License.

Monodisperse latexes characterized by small, uniform, and precisely controlled diameters of particles are highly valuable as size material references for the calibration of particle sizing equipment. Typically, the particles are composed of polystyrene and may be cross-linked with divinylbenzene to enhance their chemical resistance (Goossens et al., 1994).

7.7.2. Nanofabrication and Optoelectronics

The field of nanofabrication, which involves the creation of structures with dimensions of fewer than a hundred nanometers, is gaining momentum due to the growing demand for smaller and more efficient electronic devices. Nanostructures with regular patterns exhibit distinct optical and electric characteristics in spectral regions that are presently beyond the reach of existing technologies.

The utilization of nanostructures holds significant potential in the development of upcoming optoelectronic devices, including sensors and screens. The production of nanostructured materials can be achieved by utilizing latex particles as designs. Upon undergoing processes such as drying, settling, or filtration, polymer particles of nanoscale dimensions exhibiting uniform diameters have been observed to self-organize into ordered arrays with crystalline characteristics.

A nanoporous ordered structure can be created by incorporating a matrix material, such as a metallic or inorganic substance, into the interstices between particles. Subsequently, the polymer template can be eliminated through heating or dissolution with solvent. Latexes with conductive properties have been developed for use in electronics (Chen & Pepin, 2001).

7.7.3. Biomedical Technology

Polymer latex particles are typically considered non-toxic within the human body and can be modified to enhance their ability to interact favorably with biological systems.

Latexes possess numerous potential biomedical applications. Emulsions exhibit potential significance in controlled-release drug delivery, as suggested by previous research. Emulsion polymers possess significant interfacial areas, making them a promising option for utilization as supports in biocatalysis applications that operate at relatively low temperatures (Figure 7.13).

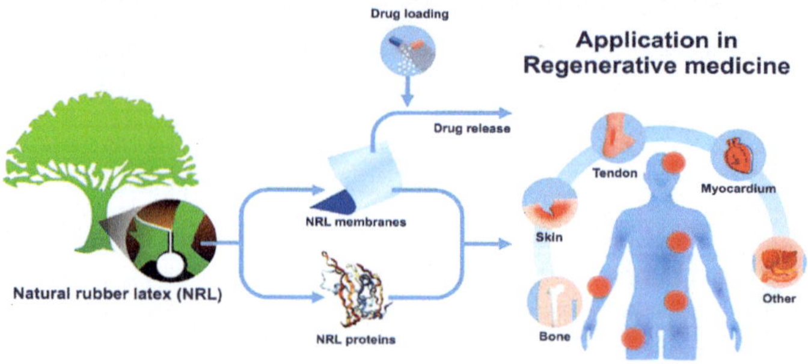

Figure 7.13. Biomedical applications of natural rubber latex from the rubber tree Hevea Brasiliensis.

Source: Nayrim Brizuela Guerra, Creative Commons License.

Latex particles that have been surface-functionalized offer a substantial interfacial area that enables biological materials such as proteins, antigens, and antibodies to interact through absorption and covalent bonds. Latex particles have been employed in medical testing immunoassay kits, commonly known as latex agglutination assays, to identify the existence of antibodies generated due to a viral infection.

The particles are covered with specific antigens and affixed to the particle's surface, which functions as agents that elicit an immune response, such as viruses and bacteria.

Upon exposure of antigens to antibodies, a bonding interaction occurs, leading to the flocculation of latex particles. The turbidity of the latex is observed to be elevated, and this alteration is identified either through visual observation or via instrumentation as an indicator of the existence of a particular antibody (Marcatto et al., 2022).

Activity 7.1

A construction company is looking for an eco-friendly and sustainable alternative for adhesive applications in their projects. They want a product with good adhesion, flexibility, and water resistance. The company aims to reduce its environmental impact while maintaining high-quality standards. How can latex-based adhesives be a suitable solution for their requirements? Discuss the advantages of latex-based adhesives in terms of sustainability and performance for construction applications.

7.8. SUMMARY

The present chapter comprehensively surveys the manifold industrial uses of latex materials. The distinctive characteristics and capabilities of latex materials have resulted in their widespread application in diverse industries. The chapter underscores the significance of latex materials in various industries, including but not limited to automobiles, medical care, construction, coverings, and textiles. The chapter provides a comprehensive analysis of the various applications of latex materials. Latex plays a critical role in the automotive sector by facilitating the production of vehicle tires, thereby improving their safety and performance. Latex gloves are essential in medical and industrial environments, as they offer exceptional barrier properties and safeguard against physiological and chemical hazards. Latex-based paints are extensively utilized in architecture and decoration due to their ease of application, rapid drying, and long-lasting nature. Using latex binders in industrial coatings confers advantages such as safeguarding metals, wood coatings, and anti-corrosion coatings. Latex coatings are employed by the paper and cardboard boxes industry to ameliorate surface characteristics, thereby augmenting printing capability and water resistance. The use of latex materials is widespread in the textiles and nonwoven sectors, as they significantly enhance fabric stability, sealing, and dimensional integrity in nonwoven fabrics. Utilizing latex materials in modifying concrete and asphalt is advantageous for the building and construction industry as it enhances the strength, durability, and crack resistance of the materials. Latex materials are used as glue binders in various applications, including carpet backing and other adhesive applications, due to their bonding properties and stability. The chapter delves into the various specialized and possible uses of latex materials. These include their emerging significance in nanofabrication, optics, and biomedical technology. The present chapter overviews the diverse industrial applications that utilize latex materials. The multifaceted nature, robustness, and efficacy-boosting attributes of latex materials render them indispensable constituents in many industries, augmenting the reliability, performance, and durability of many goods and applications.

REVIEW QUESTIONS

1. What are some industrial applications of latexes mentioned in Chapter 7?
2. Why is latex commonly used in medical gloves and condoms?

3. What are the important characteristics of latexes for textile and non-woven applications?
4. How are latexes used in building and construction, specifically in concrete and asphalt modification?
5. What are some potential specialized applications of latexes mentioned in Chapter 7?
6. What factors should be considered when selecting latexes for industrial coating applications?
7. What advantages do latex-based adhesives and sealants offer in building and construction?

REFERENCES

1. Allmers, H., Schmengler, J., & Skudlik, C., (2002). Primary prevention of natural rubber latex allergy in the German health care system through education and intervention. Journal of Allergy and Clinical Immunology, 110(2), 318–323.
2. Anedda, J., Ferreli, C., Rongioletti, F., & Atzori, L., (2020). Changing gears: Medical gloves in the era of coronavirus disease 2019 pandemic. Clinics in Dermatology, 38(6), 734–736.
3. Avramenko, V., Bratskaya, S., Zheleznov, V., Sheveleva, I., Voitenko, O., & Sergienko, V., (2011). Colloid stable sorbents for cesium removal: Preparation and application of latex particles functionalized with transition metals ferrocyanides. Journal of Hazardous Materials, 186(2, 3), 1343–1350.
4. Awoyera, P. O., Althoey, F., Ajinomisan, H., Othuman, M. M. A., Bheel, N., Sabri, S. M. M., & Ahmad, M., (2023). Potential of natural rubber latex in cement mortar for thermal insulating material in buildings. Frontiers in Materials, 10, 1152492.
5. Azahar, N. F. B. M., Hassan, N. A., Jaya, R. P., Kadir, M. A. B. A., Yunus, N. Z. B. M., & Mahmud, M. Z. H., (2016). An overview on natural rubber application for asphalt modification. Int. J. Agric. For. Plant, 2, 212–218.
6. Baldi, S., Mounayer, C., Piotin, M., Spelle, L., & Moret, J., (2003). Balloon-assisted coil placement in wide-neck bifurcation aneurysms by use of a new, compliant balloon microcatheter. American Journal of Neuroradiology, 24(6), 1222–1225.
7. Bonsignore, P. V., (1979). Flame retardant flexible polyurethane foam by post-treatment with alumina trihydrate/latex binder dispersion systems. Journal of Cellular Plastics, 15(3), 163–179.
8. Butler, L. N., Fellows, C. M., & Gilbert, R. G., (2005). Effect of surfactants used for binder synthesis on the properties of latex paints. Progress in Organic Coatings, 53(2), 112–118.
9. Callahan, M., Mauck, C., Taylor, D., Frezieres, R., Walsh, T., & Martens, M., (2000). Comparative evaluation of three TactylonTM condoms and a latex condom during vaginal intercourse: Breakage and slippage. Contraception, 61(3), 205–215.
10. Carré, M. J., Tan, S. K., Mylon, P. T., & Lewis, R., (2017). Influence of medical gloves on finger pad friction and feel. Wear, 376, 324–328.
11. Chang, J. C., Fortmann, R., Roache, N., & Lao, H. C., (1999). Evaluation of low - VOC latex paints. Indoor air, 9(4), 253–258.

12. Chen, Y., & Pepin, A., (2001). Nanofabrication: Conventional and nonconventional methods. Electrophoresis, 22(2), 187–207.
13. Cottenden, A. M., (1988). Incontinence pads and appliances. International Disability Studies, 10(1), 44–47.
14. Counts, J., Weisbrod, A., & Yin, S., (2017). Common diaper ingredient questions: Modern disposable diaper materials are safe and extensively tested. Clinical Pediatrics, 56(5_suppl), 23S–27S.
15. Crippa, M., Belleri, L., Mistrello, G., Carsana, T., Neri, G., & Alessio, L., (2003). Prevention of latex allergy among health care workers: Evaluation of the extractable latex protein content in different types of medical gloves. American Journal of Industrial Medicine, 44(1), 24–31.
16. Dillard, S. F., Hefflin, B., Kaczmarek, R. G., Petsonk, E. L., & Gross, T. P., (2002). Health effects associated with medical glove use. AORN Journal, 76(1), 88–96.
17. Dingwall, L., (2008). Promoting social continence using incontinence management products. British Journal of Nursing, 17(Sup4), S12–S19.
18. Douglas, A., Simon, T. R., & Goddard, M., (1997). Barrier durability of latex and vinyl medical gloves in clinical settings. American Industrial Hygiene Association Journal, 58(9), 672–676.
19. Edlich, R. F., Long, III. W. B., Gubler, D. K., Rodeheaver, G. T., Thacker, J. G., Borel, L., & Zura, R. D., (2009). Dangers of cornstarch powder on medical gloves: Seeking a solution. Annals of Plastic Surgery, 63(1), 111–115.
20. Fehr, S. K., Vidourek, R. A., & King, K. A., (2015). Intra-and inter-personal barriers to condom use among college students: A review of the literature. Sexuality & Culture, 19, 103–121.
21. Filon, F. L., & Radman, G., (2006). Latex allergy: A follow up study of 1040 healthcare workers. Occupational and Environmental Medicine, 63(2), 121–125.
22. Finger, W. R., (1998). Do condom characteristics influence use? Network (Vol. 18, No. 3, p. 23). Research Triangle Park, NC.
23. Goossens, M., Mittelbach, F., & Samarin, A., (1994). The LATEX Companion (Vol. 1). Reading: Addison-Wesley.
24. Grajales, III. F. J., Sheps, S., Ho, K., Novak-Lauscher, H., & Eysenbach, G., (2014). Social media: A review and tutorial of applications in medicine and health care. Journal of Medical Internet Research, 16(2), e2912.
25. Grimley, D. M., Prochaska, G. E., & Prochaska, J. O., (1997). Condom use adoption and continuation: A transtheoretical approach. Health Education Research, 12(1), 61–75.

26. Henson, W. A., Taber, D. A., & Bradford, E. B., (1953). Mechanism of film formation of latex paint. Industrial & Engineering Chemistry, 45(4), 735–739.

27. Husain, S. Z., Malik, R. N., Javaid, M., & Bibi, S., (2008). Ethonobotanical properties and uses of medicinal plants of Morgah biodiversity park, Rawalpindi. Pak. J. Bot., 40(5), 1897–1911.

28. Isozumi, H., Horiba, T., Kubota, K., Hida, K., Matsuyama, T., Yasuno, S., & Komaba, S., (2020). Application of modified styrene-butadiene-rubber-based latex binder to high-voltage operating $LiCoO_2$ composite electrodes for lithium-ion batteries. Journal of Power Sources, 468, 228332.

29. Jafarnezhadgero, A., Fakhri, E., Jafarnezhadgero, F., & Noorian, E., (2022). Designing a new anti microbe and virus volleyball ball. Journal of Advanced Sport Technology, 6(1), 31–38.

30. Järnström, J., Väisänen, M., Lehto, R., Jäsberg, A., Timonen, J., & Peltonen, J., (2010). Effect of latex on surface structure and wetting of pigment coatings. Colloids and Surfaces A: Physicochemical and Engineering Aspects, 353(2, 3), 104–116.

31. Jones, R. T., Scheppmann, D. L., Heilman, D. K., & Yunginger, J. W., (1994). Prospective study of extractable latex allergen contents of disposable medical gloves. Annals of Allergy, 73(4), 321–325.

32. Karthik, T., & Rathinamoorthy, R., (2017). Nonwovens: Process, Structure, Properties, and Applications. WPI Publishing.

33. Kluemper, M., Uhl, T., & Hazelrigg, H., (2006). Effect of stretching and strengthening shoulder muscles on forward shoulder posture in competitive swimmers. Journal of Sport Rehabilitation, 15(1), 58.

34. Knudsen, B. B., Hametner, C., Seycek, O., Heese, A., Koch, H. U., & Peters, K. P., (2000). Allergologically relevant rubber accelerators in single - use medical gloves. Contact Dermatitis, 43(1), 9–15.

35. Korniewicz, D. M., El-Masri, M., Broyles, J. M., Martin, C. D., & O'connell, K. P., (2002). Performance of latex and nonlatex medical examination gloves during simulated use. American Journal of Infection Control, 30(2), 133–138.

36. Kostansek, E., (2007). Controlling particle dispersion in latex paints containing associative thickeners. Journal of Coatings Technology and Research, 4, 375–388.

37. Kramer, A., & Assadian, O., (2016). Indications and the requirements for single-use medical gloves. GMS Hygiene and Infection Control, 11.

38. Kristoffersson, A., & Carlström, E., (1997). Tape casting of alumina in water with an acrylic latex binder. Journal of the European Ceramic Society, 17(2, 3), 289–297.

39. Kristoffersson, A., Lapasin, R., & Galassi, C., (1998). Study of interactions between polyelectrolyte dispersants, alumina and latex binders by rheological characterization. Journal of the European Ceramic Society, 18(14), 2133–2140.
40. Kumar, R. P., (2012). Latex allergy in clinical practice. Indian Journal of Dermatology, 57(1), 66.
41. Kusumawati, M., Hidayat, I. I., Haqiyah, A., & Subarno, L., (2022). The effects of forearm resistance band training on the throwing velocity of male handball players. Journal of Physical Education and Sport, 22(11), 2879–2885.
42. Laad, M. S., (2020). Polymers in sports. In: Polymer Science and Innovative Applications (pp. 485–523). Elsevier.
43. Landeck, L., Gonzalez, E., & Koch, O. M., (2015). Handling chemotherapy drugs—Do medical gloves really protect? International Journal of Cancer, 137(8), 1800–1805.
44. Le Roux, Y., Hhabe, E., Sainte-Beuve, J., Nkengafac, J., Nkeng, J., Ngolemasango, F. A., & Gobina, S., (2000). Seasonal and clonal variations in the latex and raw rubber of Hevea Brasiliensis. Journal Of Rubber Research-Kuala Lumpur-Rubber Research Institute of Malaysia-, 3(3), 142–156.
45. Lee, D. I., (1997). Latex applications in paper coating. Polymeric Dispersions: Principles and Applications, 497–513.
46. Lepoutre, P., (1989). The structure of paper coatings: An update. Progress in Organic Coatings, 17(2), 89–106.
47. Li, C. H., Mak, S. L., Tang, W. F., Wu, M. Y., & Lam, S. K., (2020). Development of IoT-based smart recycling machine to collect the wasted non-woven fabric face mask (NFM). In: 2020 IEEE International Symposium on Product Compliance Engineering-Asia (ISPCE-CN) (pp. 1–5). IEEE.
48. Li, H., Qi, Y., Zhao, Y., Chi, J., & Cheng, S., (2019). Starch and its derivatives for paper coatings: A review. Progress in Organic Coatings, 135, 213–227.
49. Liu, R., Winnik, M. A., Stefano, F. D., & Vanketessan, J., (2001). Interdiffusion vs cross-linking rates in isobutoxyacrylamide-containing latex coatings. Macromolecules, 34(21), 7306–7314.
50. Marcatto, V. A., Sant'Ana, P. G., Barbosa, G. F., Herculano, R. D., & Guerra, N. B., (2022). 3D printed - polylactic acid scaffolds coated with natural rubber latex for biomedical application. Journal of Applied Polymer Science, 139(9), 51728.

51. Mason, H., & Rosa, I., (2022). Latex allergens in the ear straps of face coverings and masks in the UK: Initial findings. COVID, 2(7), 907–915.
52. McDonald, C. J., & Devon, M. J., (2002). Hollow latex particles: Synthesis and applications. Advances in Colloid and Interface Science, 99(3), 181–213.
53. Mellström, G. A., & Boman, A., (2000). Protective gloves. Handbook of Occupational Dermatology, 417–425.
54. Mylon, P., Lewis, R., Carré, M. J., & Martin, N., (2016). An evaluation of dexterity and cutaneous sensibility tests for use with medical gloves. Proceedings of the Institution of Mechanical Engineers, Part C: Journal of Mechanical Engineering Science, 230(16), 2896–2912.
55. Mylon, P., Lewis, R., Carré, M. J., Martin, N., & Brown, S., (2014). A study of clinicians' views on medical gloves and their effect on manual performance. American Journal of Infection Control, 42(1), 48–54.
56. Najafi, S. M. H., Tajvidi, M., & Bousfield, D. W., (2018). Production and mechanical characterization of free-standing pigmented paper coating layers with latex and starch as binder. Progress in Organic Coatings, 123, 138–145.
57. Oh, K., Abhari, A. R., Im, W., Lee, J. H., Shen, Z., Kwon, S., & Lee, H. L., (2019). Effect of core-shell structure latex on pigment coating properties. BioResources, 14(1), 1241–1251.
58. Ouederni, M., (2016). Polyolefins in textiles and nonwovens. Polyolefin Compounds and Materials: Fundamentals and Industrial Applications, 231–245.
59. Palosuo, T., Antoniadou, I., Gottrup, F., & Phillips, P., (2011). Latex medical gloves: Time for a reappraisal. International Archives of Allergy and Immunology, 156(3), 234–246.
60. Paques, F. W., & Macedo, G. A., (2006). Plant lipases from latex: Properties and industrial applications. Quimica Nova, 29, 93–99.
61. Peula-García, J. M., Molina-Bolivar, J. A., Velasco, J., Rojas, A., & Galisteo-González, F., (2002). Interaction of bacterial endotoxine (lipopolysaccharide) with latex particles: Application to latex agglutination immunoassays. Journal of Colloid and Interface Science, 245(2), 230–236.
62. Pichot, C., Delair, T., & Kawaguchi, H., (2013). Specialty applications of latex polymers. Chemistry and Technology of Emulsion Polymerization, 283–305.
63. Pitten, F. A., Herdemann, G., & Kramer, A., (2000). The integrity of latex gloves in clinical dental practice. Infection, 28, 388–392.

64. Preece, D., Lewis, R., & Carré, M. J., (2021). A critical review of the assessment of medical gloves. Tribology-Materials, Surfaces & Interfaces, 15(1), 10–19.
65. Purdy, A. T., (1983). Developments in non-woven fabrics. Textile Progress, 12(4), 1–86.
66. Rajkuberan, C., Sudha, K., Sathishkumar, G., & Sivaramakrishnan, S., (2015). Antibacterial and cytotoxic potential of silver nanoparticles synthesized using latex of Calotropis gigantea L. Spectrochimica Acta Part A: Molecular and Biomolecular Spectroscopy, 136, 924–930.
67. Rannie, K., & Craig, D. M., (1997). Adolescent females' attitudes, subjective norms, perceived behavioral control, and intentions to use latex condoms. Public Health Nursing, 14(1), 51–57.
68. Raulf, M., (2014). The latex story. In: History of Allergy (Vol. 100, pp. 248–255). Karger Publishers.
69. Rego, A., & Roley, L., (1999). In-use barrier integrity of gloves: Latex and nitrile superior to vinyl. American Journal of Infection Control, 27(5), 405–410.
70. Russell-Fell, R. W., (2000). Avoiding problems: Evidence-based selection of medical gloves. British Journal of Nursing, 9(3), 139–146.
71. Sakthivel, S., & Ramachandran, T., (2012). Thermal conductivity of non-woven materials using reclaimed fibers. International Journal of Engineering Research and Applications (IJERA), 2(3), 2986.
72. Santos, A. S., Ferreira, P. J. T., & Maloney, T., (2021). Bio-based materials for nonwovens. Cellulose, 28(14), 8939–8969.
73. Sebestyen, A., (1997). Grown up ladies' bathing caps. MAKE: The Magazine of Women's Art, (75), 27–28.
74. Sharma, A., (2013). Retreading of tires. International Journal of Engineering and Advanced Technology (IJEAT), 2(6).
75. Singer, J. M., & Plotz, C. M., (1956). The latex fixation test: I. Application to the serologic diagnosis of rheumatoid arthritis. The American Journal of Medicine, 21(6), 888–892.
76. Soubam, T., & Gupta, A., (2021). Eco-friendly natural rubber latex and modified starch-based adhesive for wood-based panels application: A review. Maejo International Journal of Energy and Environmental Communication, 3(1), 49–53.
77. Spaner, D., Dolovich, J., Tarlo, S., Sussman, G., & Buttoo, K., (1989). Hypersensitivity to natural latex. Journal of Allergy and Clinical Immunology, 83(6), 1135–1137.

78. Steiner, M. J., Dominik, R., Rountree, R. W., Nanda, K., & Dorflinger, L. J., (2003). Contraceptive effectiveness of a polyurethane condom and a latex condom: A randomized controlled trial. Obstetrics & Gynecology, 101(3), 539 547.

79. Stuart, B., Guan, J., Collins, S., Thomas, P., & Ueland, M., (2022). A preliminary study of non-woven fabrics for forensic identification purposes. Australian Journal of Forensic Sciences, 1–10.

80. Swanson, M. C., Bubak, M. E., Hunt, L. W., Yunginger, J. W., Warner, M. A., & Reed, C. E., (1994). Quantification of occupational latex aeroallergens in a medical center. Journal of Allergy and Clinical Immunology, 94(3), 445–451.

81. Tarlo, S. M., Sussman, G., Contala, A., & Swanson, M. C., (1994). Control of airborne latex by use of powder-free latex gloves. Journal of Allergy and Clinical Immunology, 93(6), 985–989.

82. Taylor, J. S., & Erkek, E., (2004). Latex allergy: Diagnosis and management. Dermatologic Therapy, 17(4), 289–301.

83. Turjanmaa, K., (1987). Incidence of immediate allergy to latex gloves in hospital personnel. Contact Dermatitis, 17(5), 270–275.

84. Urban, D., Schuler, B., & Schmidt - Thümmes, J., (2013). Large volume applications of latex polymers. Chemistry and Technology of Emulsion Polymerization, 253–282.

85. Warson, H., & Finch, C. A., (2001). Applications of Synthetic Resin Latices, Latices in Diverse Applications (Vol. 3). John Wiley & Sons.

86. Wen, Y., Wang, Y., Zhao, K., & Sumalee, A., (2017). The use of natural rubber latex as a renewable and sustainable modifier of asphalt binder. International Journal of Pavement Engineering, 18(6), 547–559.

87. Winnik, M. A., & Feng, J., (1996). Latex blends: An approach to zero VOC coatings. Journal of Coatings Technology, 68.

88. Woolf, S. E., & Maisto, S. A., (2008). Gender differences in condom use behavior? The role of power and partner-type. Sex Roles, 58, 689–701.

89. Yabuuchi, N., Kinoshita, Y., Misaki, K., Matsuyama, T., & Komaba, S., (2015). Electrochemical properties of $LiCoO_2$ electrodes with latex binders on high-voltage exposure. Journal of the Electrochemical Society, 162(4), A538.

90. Yildirim, Y., (2007). Polymer modified asphalt binders. Construction and Building Materials, 21(1), 66–72.

91. Yip, E., & Cacioli, P., (2002). The manufacture of gloves from natural rubber latex. Journal of Allergy and Clinical Immunology, 110(2), S3–S14.

92. Yunginger, J. W., Jones, R. T., Fransway, A. F., Kelso, J. M., Warner, M. A., & Hunt, L. W., (1994). Extractable latex allergens and proteins in disposable medical gloves and other rubber products. Journal of Allergy and Clinical Immunology, 93(5), 836–842.
93. Zare, A., Choobineh, A., Jahangiri, M., & Malakoutikhah, M., (2021). How do medical gloves affect manual performance? Evaluation of ergonomic indicators. International Journal of Industrial Ergonomics, 81, 103062.
94. Zare, A., Choobineh, A., Mokarami, H., & Jahangiri, M., (2021). The medical gloves assessment tool (MGAT): Developing and validating a quantitative tool for assessing the safety and ergonomic features related to medical gloves. Journal of Nursing Management, 29(3), 591–601.

CHAPTER

8 Future Trends and Troubleshooting in Latex Bonding

Contents

8.1. INTRODUCTION

Staying abreast of the most recent trends and advancements is imperative in the constantly evolving domain of latex bonding. The ability to comprehend and adjust to developing patterns can have a substantial influence on the efficacy, productivity, and competitiveness of adhesive procedures and goods involving latex. Keeping up with the latest advancements enables researchers, engineers, and business experts to utilize state-of-the-art technologies and methodologies, resulting in better bonding efficacy, heightened product characteristics, and increased potential for diverse applications. The development of novel techniques for bonding latex is influenced by a range of factors, such as progress in the field of material science, growing environmental consciousness, market requirements, and advancements in technology. Through the active pursuit and implementation of these emerging patterns, professionals can acquire a strategic advantage, explore novel avenues, and tackle obstacles within the domain of latex adhesion (Momayez et al., 2005).

The objective of the chapter pertaining to "Future Trends in Latex Bonding" is to offer discernment into the most recent advancements and nascent patterns that are influencing the trajectory of the industry. The text delves into cutting-edge technologies, original adhesive compositions, prevailing market patterns, methods for resolving issues, and prospective domains for additional research and advancement. The chapter has been segmented into multiple primary sections. Recent Developments in Latex Bonding Technology: The present section provides an overview of the most recent developments in the field of latex bonding technology. The present study investigates the advancement of innovative adhesive compositions, encompassing high-efficiency and environmentally sustainable alternatives. The text underscores the incorporation of nanotechnology into the process of latex bonding, which facilitates meticulous regulation of bond characteristics and the exploitation of distinctive nanostructures. Furthermore, the section explores the progress made in surface modification methodologies aimed at improving adhesion, including plasma treatment as well as functional coatings (Zatirakha et al., 2016).

This chapter delves into the increasing market trends and potential applications of latex-based adhesives across diverse industries. This paper investigates the increasing utilization of latex bonding within the healthcare industry, encompassing the production of medical devices and wound care commodities. The section further elaborates on the current developments in the automotive as well as construction sectors, wherein the application of latex bonding is gaining traction for purposes such as light weighting, noise abatement, and structural enhancements. The aforementioned section offers valuable perspectives on the market forecasts and expansion prospects concerning latex bonding. This analysis takes into account various factors such as globalization, sustainability, and evolving consumer inclinations (Takeyama & Matsui, 1969).

The present section is dedicated to outlining strategies for troubleshooting common issues encountered during latex bonding, with the aim of effectively addressing them. The document offers direction pertaining to the identification and mitigation of bonding failures, encompassing tactics for addressing bonding anomalies and areas of reduced strength. The segment additionally discusses difficulties pertaining to substrate compatibility as well as surface preparation, providing pragmatic remedies for surmounting these concerns.

Furthermore, the article examines the difficulties encountered in latex bonding processes on a large scale, taking into account variables such as equipment enhancement and process management. The concluding segment of the chapter delves into potential avenues for future research and opportunities in the field of latex bonding.

The text highlights crucial domains that require additional investigation and advancement, such as the examination of innovative substances and supplements to enhance bonding efficacy. The segment additionally deliberates on progressions in the characterization and testing methodologies for assessing bond quality, encompassing non-invasive testing approaches and sophisticated analytical instruments.

Moreover, it underscores the significance of cooperation and information exchange among scholars, experts in the industry, and academic institutions to propel the progress of latex bonding technology.

The chapter on "Future Trends in Latex Bonding" endeavors to offer a thorough exposition of the significance of remaining abreast of nascent trends and their possible ramifications on the domain of latex bonding by addressing these subjects.

This resource is highly valuable for professionals in the industry, researchers, and students who aim to navigate the dynamic landscape of latex bonding as well as foster progress in this area through innovative approaches (Debuigne et al., 2009).

8.2. ADVANCEMENTS IN LATEX BONDING TECHNOLOGY

The domain of adhesive bonding has experienced a substantial metamorphosis owing to the progressions in latex bonding technology, leading to improved efficacy, versatility, and environmental sustainability. An important advancement in the field concerns the development of novel adhesive compositions and techniques. The formulations mentioned above exhibit self-healing properties that enable autonomous repair of small imperfections and microcracks, thereby enhancing the overall durability and lifespan of adhesively bonded products. Another significant advancement concerns the integration of eco-friendly and sustainable techniques in latex bonding.

The utilization of bio-based and renewable materials as binding agents is a feasible approach to reduce reliance on non-renewable resources and alleviate ecological ramifications. The integration of nanoparticles into the latex matrix has yielded notable enhancements in latex adhesion, leading to augmented robustness, potency, and resistance to impact. The field of nanotechnology has played a crucial role in enabling these progressions. Moreover, advancements in surface modification techniques have improved adhesion by optimizing surface interactions and increasing wetting abilities. The advancements in latex bonding technology have opened up new possibilities for diverse applications, leading to the creation of adhesive bonding alternatives that possess greater durability, longevity, and environmental sustainability (Bashir et al., 2022).

8.2.1. Novel Adhesive Formulations and Approaches

The domain of latex bonding has experienced noteworthy progressions in adhesive compositions and methodologies. Innovative methods to enhance bond performance, adaptability, and efficiency have been under exploration by both researchers and industry professionals. Advanced adhesive compositions provide improved bonding efficacy, superior adhesion to a wide range of substrates, and customized characteristics to suit particular use cases. In contemporary times, there has been a transition towards the creation of multifaceted adhesives that amalgamate robust bonding capabilities with supplementary functionalities. Adhesive formulations possessing self-healing properties have garnered significant attention. The application of these adhesives can result in autonomous repair of minor defects or microcracks, thereby enhancing the overall durability as well as durability of bonded products. An additional field of investigation pertains to the advancement of adhesives that are inspired by biological systems. Scholars have sought inspiration from naturally occurring adhesive mechanisms, such as those found in gecko feet or mussel adhesives, in order to fabricate artificial adhesives that exhibit remarkable adhesion characteristics. The bio-inspired adhesives exhibit robust adhesion on diverse substrates and hold the potential to transform the field of latex bonding in demanding scenarios (Fan et al., 2016) (Figure 8.1).

Tip

Stay updated with the latest developments in adhesive formulations and approaches for latex bonding to leverage innovative solutions that can enhance bonding strength and performance.

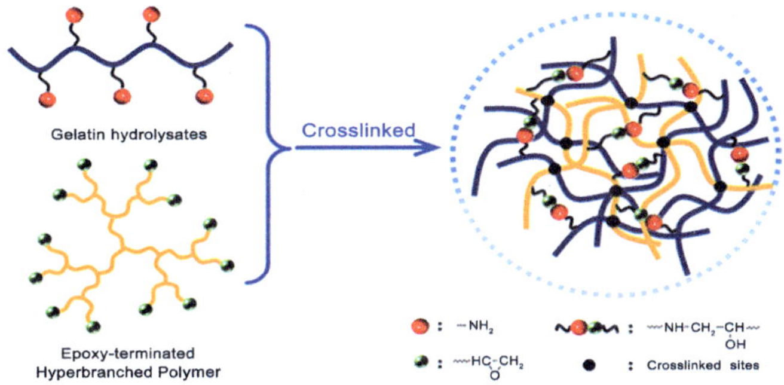

Figure 8.1. Novel gelatin-based eco-friendly adhesive with a hyperbranched cross-linked structure.

Source: Xuechuan Wang, Creative Commons License.

8.2.2. Development of Eco-friendly Latex Bonding Solutions

The emergence of sustainable as well as environmentally friendly latex bonding solutions can be attributed to the growing environmental concerns. The industrial sector is currently exploring substitutes for conventional adhesive systems that potentially comprise of dangerous chemicals or produce elevated quantities of volatile organic compounds (VOCs). Consequently, there has been a notable emphasis on the advancement of latex adhesives with low levels of VOCs or those that are entirely devoid of VOCs. The utilization of bio-derived and sustainable substances in the process of latex adhesion has garnered significant attention in recent times. Scholars are currently investigating the potential utilization of naturally occurring polymers obtained from sustainable origins, like starch, cellulose, and proteins, as binding agents in adhesive compositions. Bio-based adhesives provide ecologically sustainable options while upholding satisfactory bonding efficacy.

In addition, endeavors are being undertaken to enhance the recyclable and reusable properties of products bonded with latex. Research is currently being conducted to investigate advancements in adhesive technologies that facilitate the effortless separation as well as recycling of bonded materials, while maintaining the integrity of the bond strength. The aforementioned practice advocates for a sustainable methodology in the realm of latex bonding, thereby mitigating the generation of waste and minimizing the utilization of resources (Hidayat et al., 2022).

8.2.3. Integration of Nanotechnology in Latex Bonding

The incorporation of nanotechnology has presented novel prospects in the realm of latex adhesion. Nanoparticles and nanomaterials possess distinctive characteristics and functionalities that have the potential to augment bond performance and customize adhesive properties.

An intriguing field of study pertains to the integration of nanoparticles to latex adhesive compositions with the aim of enhancing their mechanical characteristics and adhesive capabilities.

The incorporation of nanoparticles, like silica, carbon nanotubes, and graphene, into the latex matrix can lead to a reinforcement effect, which can ultimately result in improved durability, strength, and resistance to impact of the bond. In addition, the surface alteration of nanoparticles has the potential to enhance their dispersion as well as integration in the latex creation. The field of nanotechnology has been utilized to create adhesives that are responsive to stimuli.

The reversible bonding properties of adhesives can be achieved through the integration of responsive nanomaterials, like shape-memory polymers and temperature-sensitive nanoparticles, which enable the adhesives to exhibit temporary or repositionable bonding capabilities. The intelligent adhesives exhibit a capacity to react to external stimuli, thereby facilitating novel functionalities in the realm of latex bonding applications (Verd et al., 2005) (Figure 8.2).

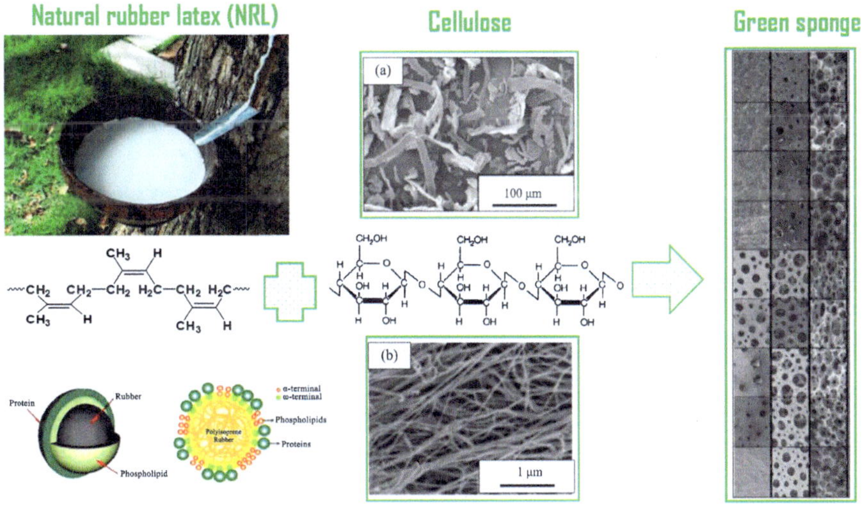

Figure 8.2. Natural rubber latex foam reinforced with micro- and nanofibrillated cellulose via Dunlop method.

Source: Sirilak Phomrak, Creative Commons License.

8.2.4. Advancements in Surface Modification Techniques

The field of surface modification techniques has experienced notable progress in augmenting the adhesion properties of latex bonding. The objective of these techniques is to enhance the adhesive strength and longevity through the modification of substrate surface properties, which facilitates superior wetting and augments the interfacial contact region. The utilization of plasma treatment as a surface alteration method in the process of latex bonding has become increasingly prevalent. The application of plasma treatment induces alterations in the chemical composition and surface energy of substrates, thereby facilitating improved adhesion and wetting characteristics. This technique has the ability to turn on or functionalize diverse substrates such as polymers, metals, ceramics, as well as composites, thereby improving their bonding performance (Luo & Van Ooij, 2002).

An additional methodology involves the utilization of useful coatings or primers applied onto the surfaces of substrates. The aforementioned coatings offer a customized interfacial stratum that enhances the adherence and congruity amidst the latex as well as the substrate. Functional coatings have the potential to augment wetting properties, amplify surface roughness, or furnish chemical functionalities that facilitate robust interfacial bonding. Furthermore, the progress made in surface analysis and description methodologies has enabled a more comprehensive comprehension and regulation of interfacial properties. The utilization of advanced analytical methods like atomic force microscopy (AFM), X-ray photoelectron spectroscopy (XPS), as well as contact angle measurements facilitates comprehensive surface characterization, thereby facilitating the refinement of surface modification techniques and enhancing the comprehension of the adhesion mechanisms involved in latex bonding. To conclude, the progress made in the field of latex bonding technology has facilitated the development of innovative adhesive compositions, environmentally sustainable alternatives, the incorporation of nanotechnology, and techniques for modifying surfaces. The aforementioned developments provide enhanced adhesive potency, adaptability, ecological soundness, and customized characteristics to cater to the changing requirements of diverse uses. The integration of these developments in latex bonding has the potential to yield improved performance, durability, as well as competitiveness within the industry (Remanan et al., 2018).

8.3. POTENTIAL APPLICATIONS AND MARKET TRENDS

The utilization of latex-based adhesives is consistently increasing in diverse sectors owing to their multifunctionality, efficacy, and ecological benefits. Adhesives based on Latex exhibit distinctive characteristics such as pliability, elevated adhesive potency, and resilience against dampness and thermal fluctuations. Consequently, they are extensively employed in a diverse range of applications, spanning from consumer goods to industrial settings (Mabrouk et al., 2020). The need for latex-based adhesives is being driven by the packaging industry, which is considered a crucial sector. The utilization of these adhesives is widespread in the sealing of cartons, lamination of films, and bonding of diverse packaging materials. Latex-based adhesives are highly regarded in the packaging industry due to their rapid bonding rate, strong adhesion to various substrates, and capacity to uphold bond integrity throughout the transportation and storage processes. The furniture as well as woodworking industry is witnessing an increasing demand for adhesives based on latex. Latex adhesives exhibit superior adhesive properties when bonding wood and wood-derived materials, rendering them highly suitable for use in the production of furniture, cabinetry, and veneer applications.

The furniture industry places significant value on the non-toxic properties of latex adhesives. Furthermore, latex-based adhesives are extensively utilized by the textiles as well as footwear sectors. The utilization of these adhesives is prevalent in the bonding of textiles, such as fabrics and leather, and other related materials. Textile materials possess several benefits including adaptability, launderability, and resilience, rendering them appropriate for various uses such as the fabrication of clothing, the manufacturing of footwear, and the creation of furniture coverings (Mapari et al., 2021).

8.3.1. Expanding Applications in the Healthcare Sector

The healthcare industry has observed a proliferation of latex bonding applications. The healthcare industry has witnessed a surge in demand for medical equipment, wound care goods, and drug delivery systems, thereby prompting a thorough investigation into the application of latex bonding within this domain. Latex-based adhesives are commonly employed in the medical device manufacturing industry to bond various components, including tubing, connectors, as well as sensors. The aforementioned adhesives offer dependable and biocompatible adhesion, thereby safeguarding the coherence and efficacy of medical apparatus. The utilization of latex bonding presents several benefits, including its pliability, ability to withstand sterilization techniques, and adaptability to diverse medical-grade substances (Bandl et al., 2020) (Figure 8.3).

Figure 8.3. Perspectives for the application of plant latex and its constituents in agriculture, medicine, and industry.

Source: *Robert Nawrot, Creative Commons License.*

Latex-based adhesives are of significant importance in wound care uses as they serve the purpose of securing medical tapes and dressings. The aforementioned adhesives exhibit a mild adhesive property to the integumentary system, thereby ensuring steadfast fixation while facilitating painless detachment without inducing discomfort or epidermal impairment. Latex adhesives' ability to allow air to pass through them enables effective moisture control and expedites the process of wound recovery.

Did You Know?

Latex bonding is finding increased applications in the healthcare sector, including medical devices, wound dressings, and surgical equipment, due to its biocompatibility, flexibility, and adhesive properties.

Furthermore, the advancement of transdermal delivery systems for drugs has propelled the utilization of latex bonding in medicine contexts. Adhesives based on Latex are utilized for the purpose of affixing drug patches or diagnostic patches. The aforementioned adhesives provide dependable adherence to the skin, regulated dispensation of medication, and enhanced adherence to medical treatment by patients (Ain & Azura, 2011).

8.3.2. Emerging Trends in Automotive and Construction Industries

The automotive as well as construction sectors are currently observing developing patterns in latex bonding, which are being propelled by the requirements for enhanced performance, sustainability, and lightweighting (Imam et al., 2013).

The automotive sector has exhibited an increasing inclination towards the utilization of latex bonding for purposes of lightweighting. Adhesives based on Latex exhibit remarkable bonding capabilities when it comes to the adhesion of lightweight materials, including composites, high-strength steels, and aluminum. Latex bonding has the potential to replace conventional welding or mechanical securing techniques, resulting in reduced weight, improved fuel efficiency, and enhanced overall performance of vehicles (Bijina et al., 2022).

Latex bonding is being increasingly utilized in diverse applications within the construction industry. Adhesives that are based on Latex are commonly utilized in the process of installing various types of flooring products, wall coverings, as well as paneling. The aforementioned adhesives exhibit robust bonding properties to various substrates, possess pliability to accommodate structural shifts, and demonstrate resilience to environmental stressors such as humidity and temperature variations. Furthermore, sealants based on latex are employed in order to seal joints and gaps, thereby guaranteeing the impermeability to air and water in construction endeavors (Patnaik et al., 2020).

8.3.3. Market Projections and Growth Opportunities

The latex bonding industry is anticipated to witness substantial expansion in the foreseeable future, owing to various factors including the proliferation of applications, advancements in technology, and the need for sustainability. The increasing need for environmentally conscious and enduring adhesive alternatives has created fresh prospects in the domain of latex bonding industry. Adhesives based on Latex exhibit characteristics that are in accordance with the growing regulations governing the environment and consumer demand for eco-friendly products, owing to their minimal VOC content, non-hazardous properties, and availability of renewable material alternatives (Bai et al., 2021).

Moreover, the progressions in latex bonding technology, encompassing innovative adhesive compositions and enhanced efficacy, are propelling the expansion of the market. Industries are increasingly interested in the creation of high-performance latex adhesives that possess customized properties to suit specific applications. This is due to the desire for dependable and effective bonding solutions. The anticipated growth of the latex bonding market is expected to be significantly influenced by new economies in Asia-Pacific as well as Latin America, as per geographical trends. The widespread use of latex-based adhesives in different industries is being fueled by swift industrialization, construction of infrastructure, and expanding demand from customers in these regions (Zhao et al., 2011).

To conclude, there is an increasing demand for adhesives based on latex in various sectors including but not limited to packaging, furnishings, textiles, medical care, automotive, as well as construction. The potential for possibilities for expansion in the latex bonding market is underscored by growing uses in healthcare, the new developments in automotive as well as construction, and market projections. The market prospects are reinforced by the trend towards environmentally friendly as well as sustainable adhesive solutions, which is stimulating innovation and the advancement of sophisticated latex bonding technologies (Ohama, 1987).

8.4. TROUBLESHOOTING COMMON CHALLENGES IN LATEX BONDING

The process of latex bonding may face a range of challenges that can potentially impact the performance of the bond, similar to other adhesive processes. The identification and resolution of these challenges are imperative in guaranteeing favorable bonding results. The present section is dedicated to addressing prevalent issues encountered during latex bonding and proposing effective approaches to tackle them.

8.4.1. Identification and Prevention of Bonding Failures

The occurrence of bonding failures can be attributed to several factors such as insufficient surface preparation, inappropriate adhesive formulation, or erroneous bonding techniques. The identification of underlying factors that contribute to bonding failures is imperative, and proactive measures must be taken to prevent such occurrences. Inadequate surface preparation is a prevalent factor contributing to bonding failure. Inadequate cleaning, contamination, or inappropriate surface roughness may impede the process of adhesive bonding. The implementation of comprehensive surface cleaning and suitable treatment techniques, like solvent wiping, rough cleaning, and plasma treatment, can effectively augment surface adhesion and forestall bonding failures (Ryan et al., 2014).

Inadequate bonding may also result from inappropriate adhesive formulation or selection. The bond may be compromised due to factors such as incompatibility among the adhesive as well as substrate, insufficient adhesive strength, and inadequate adhesive thickness. Comprehending the bonding necessities and opting for appropriate adhesives possessing the requisite characteristics is of paramount importance in order to avert bonding malfunctions (Bregar et al., 2020).

8.4.2. Strategies for Addressing Bonding Defects

Despite appropriate surface preparation as well as adhesive selection, bonding imperfections and weak spots may still arise. To tackle these concerns, it is necessary to implement specific approaches that aim to enhance the connection and enhance the overall efficiency. To address bonding defects, such as voids or gaps, it is recommended to utilize adhesive injection into the affected regions or employ pressure-assisted bonding techniques. These methods can effectively fill the gaps and improve the overall integrity of the bond. The utilization of localized heat or heat-activated adhesives has been found to enhance the formation of bonds in difficult-to-bond regions (Courard et al., 2013).

The implementation of reinforcement techniques can prove to be efficacious in enhancing areas of weakness within a bond. The incorporation of reinforcing materials, like fibers, meshes, or tapes, inside the adhesive system has been found to improve bond strength and promote more uniform stress distribution. Furthermore, the utilization of adhesive additives and modifiers that augment toughness and flexibility has the potential to enhance mechanical stress resistance and forestall localized failures (Guo et al., 2021) (Figure 8.4).

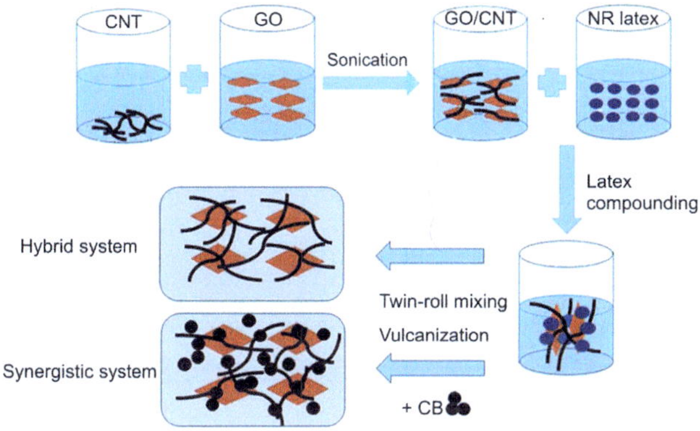

Figure 8.4. Schematic description of the latex mixing method used for the preparation of the reinforced latex composite.

Source: Laiyun Wei, Creative Commons License.

8.4.3. Dealing with Substrate Compatibility and Surface Preparation

The attainment of robust and long-lasting latex bonds is contingent upon the critical factors of substrate compatibility as well as surface preparation. Difficulties in these domains may emerge due to differences in surface energies, impurities, or intricate substrate materials. The resolution of substrate compatibility concerns necessitates a comprehensive comprehension of the surface characteristics of the substrates and the selection of adhesives that demonstrate favorable wetting and adhesion properties. The implementation of appropriate surface preparation methods that are customized to the specific substrates, including but not limited to cleaning, priming, and surface activation, can significantly improve substrate compatibility and facilitate efficient bonding. The presence of contaminants on the surfaces of substrates has the potential to undermine the adhesive strength of bonds. It is imperative to employ suitable cleaning techniques to eliminate impurities such as oils, greases, or release agents. The implementation of appropriate solvents, detergents, or surface treatment techniques, such as corona treatment or flame treatment, can proficiently eliminate impurities and enhance adhesion. Achieving interfacial compatibility can be a challenging task when bonding dissimilar or complex substrates. Enhancing interfacial compatibility and bonding performance can be achieved by utilizing adhesive systems with improved adhesion to multiple substrates or by utilizing coupling agents or primers that are specifically tailored to the substrate combinations (Morgan, 1996).

8.4.4. Troubleshooting Challenges in Large-scale Processes

The execution of latex bonding processes on a large scale presents distinctive difficulties that arise from various factors, including but not limited to adhesive drying times, equipment constraints, and process regulation. Addressing these issues necessitates a thorough comprehension and refinement of the bonding procedure. Achieving uniform drying of adhesives over extensive areas can pose a difficulty. The optimization of drying conditions, including temperature, humidity, and airflow, can facilitate homogeneous drying and avert non-uniform adhesive thickness or inadequate curing. The utilization of drying aids, like infrared heating and vacuum-assisted drying, has been found to enhance the effectiveness and homogeneity of adhesive drying (Adam et al., 2016).

Remember

When troubleshooting challenges in large-scale latex bonding processes, consider factors such as process variables, equipment maintenance, and quality control protocols to ensure consistent and reliable bonding outcomes.

The process of bonding on a large scale frequently entails intricate machinery and logistical factors to be taken into account. Achieving uniform bonding outcomes is contingent upon several critical factors, including the maintenance of appropriate equipment calibration, the assurance of consistent adhesive application, and the management of process parameters. The implementation of routine servicing of equipment, operator training, and procedure monitoring is imperative for effectively addressing any potential complications that may arise during the course of large-scale latex bonding procedures. To summarize, the resolution of typical obstacles in latex bonding necessitates the identification of underlying factors leading to bonding malfunctions, the implementation of preventative measures, and the utilization of specific approaches to rectify bonding defects and vulnerabilities. The comprehension of substrate compatibility, the optimization of surface preparation, and the adoption of appropriate bonding techniques are crucial in addressing challenges. Furthermore, addressing troubleshooting issues pertaining to latex bonding processes on a large scale necessitates the optimization of the process, maintenance of equipment, and implementation of process control measures. Through effective management of these challenges, professionals can attain dependable, robust, and enduring latex adhesions across diverse applications (Provencher et al., 2017).

8.5. FUTURE DIRECTIONS AND RESEARCH OPPORTUNITIES

The domain of latex bonding presents a plethora of prospects for additional investigation and enhancement to propel the comprehension, potentialities, and efficacy of adhesives that are based on latex. Crucial domains for future investigation encompass:

1. **Enhanced Bonding mechanisms:** The exploration and comprehension of the underlying processes of latex bonding at the molecular as well as interfacial scales can offer valuable perspectives for enhancing adhesive compositions, surface interactions, as well as interfacial adhesion. Investigating the function of distinct functional groups, interactions between molecules, as well as interfacial phenomena has the potential to facilitate the advancement of latex bonding systems that are more effective and dependable.
2. **Tailored Adhesion for Challenging Substrates:** The research area of developing adhesives that demonstrate robust adhesion to difficult substrates, including low-surface-energy substances, highly textured surfaces, and chemically resistant materials, is of considerable importance. Comprehending the surface characteristics and interfacial phenomena between adhesives and substrates can facilitate the development of customized adhesive compositions and surface alteration methodologies for augmenting adhesion efficacy.

3. **Adhesion in Extreme Environments:** The exploration of the efficacy of latex-based adhesives under severe conditions, including but not limited to elevated temperatures, excessive moisture, and harsh chemical surroundings, offers prospects for scholarly inquiry. The development of adhesives that can sustain their bond durability and strength in challenging environments is of utmost importance for their utilization in the aerospace, automotive, as well as oil and gas sectors (Imtiaz et al., 2022).

8.5.1. Exploration of Novel Materials and Additives

The investigation of new materials and additives has the potential to enhance the bonding efficacy in applications involving latex bonding. The research can be centered on:

1. **Bio-based and Renewable Materials:** The exploration of bio-based polymer compounds, natural resins, as well as renewable resources as substitute binders in latex adhesives presents a promising avenue for sustainable bonding alternatives. These materials have the potential to offer bond performance that is either equivalent or superior, while simultaneously decreasing dependence on non-renewable fossil resources.
2. **Nanomaterials for Enhanced Properties:** The investigation of the integration of nanomaterials, specifically nanoparticles, nanofibers, and nanocomposites, into latex adhesives presents the potential for enhanced mechanical characteristics, thermal endurance, and functional features. The precise manipulation of nanomaterial dispersion and alignment within the latex matrix has the potential to improve interfacial adhesion as well as overall bond performance.
3. **Smart Additives and Stimuli-Responsive Systems:** The exploration of incorporating intelligent additives, like shape-memory polymers, self-healing agents, and pH-responsive materials, into latex adhesives has the potential to facilitate improved functionality and adaptability. The incorporation of additives in materials can result in a response to external stimuli, thereby enhancing the bonding performance, repairability, and adaptability of the material in dynamic applications (Karana et al., 2015).

8.5.2. Advancements in Characterization and Testing Methods

The evaluation of bond quality in latex bonding is of paramount importance, and as such, the development of characterization and testing techniques is imperative. The research endeavors may concentrate on:

1. **Non-destructive Evaluation Techniques:** The investigation of non-destructive evaluation techniques, including acoustic methods, the optical coherence tomography, and thermal imaging, has the potential to facilitate instantaneous monitoring and evaluation of bond integrity while preserving the structural integrity of the bonded assembly. The utilization of these techniques can yield valuable information pertaining to the strength of bonds, interfacial characteristics, and possible flaws or vulnerabilities.
2. **Surface Analysis Techniques:** Sophisticated surface analysis methodologies, including excellent quality microscopy, photoelectron spectroscopy with X-rays (XPS), or time-of-flight secondary ion mass spectrometry (ToF-SIMS), have the potential to furnish comprehensive insights into interfacial interactions, surface chemistry, and surface topography. The implementation of these methods can facilitate comprehension of the interactions between adhesives and substrates, as well as the enhancement of surface preparation techniques.
3. **Mechanical Testing Methods:** The implementation of sophisticated mechanical testing techniques, including microbond testing, nanoindentation, and microscale tensile testing, can facilitate accurate assessment of bond power, adhesion properties, as well as failure mechanisms at reduced scales. The utilization of these techniques can offer valuable perspectives on the performance of bonds in specific regions and facilitate the development of adhesive systems that are more dependable (Fulghum, 1999).

8.5.3. Collaboration and Knowledge Sharing

The advancement of latex bonding technology is significantly influenced by collaboration and knowledge sharing. Collaboration among academic institutions, business partners, as well as adhesive manufacturers can be established in order to:

1. **Share Expertise and Best Practices:** Collaborative platforms and networks have the potential to enable the transfer of knowledge, experiences, as well as best practices within researchers, industry professionals, as well as adhesive formulators. The act of collaborating has the potential to expedite progress in the field of latex bonding through the exchange of knowledge, obstacles, and resolutions.
2. **Collaborative Research Projects:** Collaborative research endeavors that involve academic institutions, industrial entities, and research organizations can effectively combine resources, specialized knowledge, and infrastructure to address intricate issues related to the bonding of latex. The pooling of resources and expertise among individuals or groups can result in novel problem-solving approaches, streamlining of adhesive bonding procedures, and the emergence of fresh adhesive advancements (Saqr et al., 2018).
3. **Industry-Academia Partnerships:** The establishment of robust collaborations between academia and industry has the potential to facilitate the dissemination of knowledge, encourage the transfer of technology, and stimulate innovation in the field of latex bonding. The formation of partnerships can serve as a means to enable cooperative research endeavors, provide opportunities for student internships, and foster the joint creation of novel adhesive products and their corresponding applications.

To summarize, forthcoming investigations on latex bonding ought to concentrate on crucial domains, including improved bonding mechanisms, innovative substances and supplements, progressions in characterization and testing methodologies, and cooperation and dissemination of knowledge. The investigation of these prospects has the potential to propel progress in the domain of latex bonding technology, resulting in enhanced bonding efficacy, broadened utility, and ecologically sound adhesive resolutions (Xu & Logan, 2005).

Activity 8.1

A company is experiencing frequent bonding failures in their latex bonding processes. Despite following the recommended procedures, they are facing challenges in achieving strong and durable bonds. As a troubleshooting expert, provide guidance to address this issue. Discuss potential causes of bonding failures, strategies for prevention, and steps to troubleshoot and improve the bonding quality in large-scale latex bonding processes.

8.6. SUMMARY

The chapter pertaining to "Future Trends and Troubleshooting in Latex Bonding" offers significant perspectives on the developing domain of latex bonding and tackles the obstacles encountered in attaining the most favorable bond efficacy. The chapter is structured into five principal segments, comprising an initial exposition on forthcoming trends, developments in latex bonding technology, possible applications and market tendencies, resolution of typical difficulties, and forthcoming avenues for research and cooperation. The introductory section of the chapter emphasizes the significance of remaining informed about the latest developments in latex bonding as well as presents a summary of the chapter's contents. The statement underscores the importance of comprehending and integrating recent advancements in adhesive formulations, environmentally sustainable solutions, incorporation of nanotechnology, techniques for surface modification, and prevailing market trends.

The segment pertaining to progressions in latex bonding technology explores innovative adhesive compositions and methodologies, encompassing self-repairing adhesives and adhesives inspired by biological systems. The text delves into the investigation of environmentally conscious and enduring latex adhesion remedies, as well as the assimilation of nanotechnology in latex bonding to enhance characteristics and capabilities. Furthermore, the article delves into the progress made in surface modification methodologies aimed at improving adhesion. The subsequent chapter delves into the potential uses and market patterns in latex bonding, with a particular emphasis on the increasing need for adhesives based on latex in diverse sectors, including but not limited to packaging, furnishings, textiles, healthcare, automotive, as well as construction. The text underscores the increasing utilization of latex bonding within the healthcare industry, encompassing diverse areas such as the production of medical devices, wound care commodities, and transdermal drug administration mechanisms. Furthermore, an analysis is conducted on the developing patterns within the automotive as well as construction sectors, taking into account the objectives of reducing weight, enhancing performance, and meeting sustainability standards. The discussion encompasses market projections and development opportunities pertaining to latex bonding.

The segment pertaining to the resolution of typical obstacles encountered in latex bonding concerns the recognition and avoidance of bonding malfunctions, methods for rectifying bonding imperfections and areas of vulnerability, management of substrate compatibility as well as surface preparation predicaments, and the resolution of difficulties in latex bonding procedures on a large scale. The statement underscores the significance of comprehending the fundamental reasons behind bonding failures, executing preemptive measures, and utilizing specific tactics to augment bond efficacy and longevity. The chapter concludes by delving into potential avenues for future research and development in the field of latex bonding. The text highlights crucial domains that require additional investigation and progress, such as improved bonding mechanisms, innovative materials as well as additives, advancements in description and testing techniques, and cooperation and dissemination of knowledge. These regions offer a route for the progression of latex bonding technology, the enhancement of bond efficiency, and the fulfillment of changing industry requirements.

REVIEW QUESTIONS

1. What are some advancements in latex bonding technology discussed in Chapter 8?
2. How does the integration of nanotechnology contribute to latex bonding?
3. What are the potential applications and market trends for latex bonding in the healthcare sector?
4. What emerging trends are observed in the automotive and construction industries regarding latex bonding?
5. How can common challenges in latex bonding be identified and prevented?
6. What strategies can be employed to address bonding defects in latex bonding processes?
7. Why is substrate compatibility and surface preparation important in latex bonding?

REFERENCES

1. Adam, R., Ade, P. A. R., Alves, M. I. R., Ashdown, M., Aumont, J., Baccigalupi, C., & Santos, D., (2016). Planck intermediate results-XLII. Large-scale galactic magnetic fields. Astronomy & Astrophysics, 596, A103.
2. Ain, Z. N., & Azura, A. R., (2011). Effect of different types of filler and filler loadings on the properties of carboxylated acrylonitrile–butadiene rubber latex films. Journal of Applied Polymer Science, 119(5), 2815–2823.
3. Bai, M., Cao, J., Li, J., & Li, C., (2021). Development of soybeans starch based tough, water resistant and mildew-proof adhesives through multiple cross linking cooperation strategy. Journal of Cleaner Production, 321, 129001.
4. Bandl, C., Kern, W., & Schlögl, S., (2020). Adhesives for "debonding-on-demand": Triggered release mechanisms and typical applications. International Journal of Adhesion and Adhesives, 99, 102585.
5. Bashir, O., Bhat, S. A., Basharat, A., Qamar, M., Qamar, S. A., Bilal, M., & Iqbal, H. M., (2022). Nano-engineered materials for sensing food pollutants: Technological advancements and safety issues. Chemosphere, 292, 133320.
6. Bijina, V., Jandas, P. J., Joseph, S., Gopu, J., Abhitha, K., & John, H., (2022). Recent trends in industrial and academic developments of green tire technology. Polymer Bulletin, 7, 1–30.
7. Bregar, T., An, D., Gharavian, S., Burda, M., Durazo-Cardenas, I., Thakur, V. K., & Yazdani, N. H., (2020). Carbon nanotube embedded adhesives for real-time monitoring of adhesion failure in high performance adhesively bonded joints. Scientific Reports, 10(1), 1–20.
8. Courard, L., Bissonnette, B., Beushausen, H., Fowler, D., Trevino, M., Alex, V., & Johan, S., (2013). Recommendations for the repair, the lining or the strengthening of concrete slabs or pavements with bonded cement-based material overlays. Materials and Structures, 46, 2159.
9. Debuigne, A., Poli, R., Jérôme, C., Jérôme, R., & Detrembleur, C., (2009). Overview of cobalt-mediated radical polymerization: Roots, state of the art and future prospects. Progress in Polymer Science, 34(3), 211–239.
10. Fan, B., Zhang, L., Gao, Z., Zhang, Y., Shi, J., & Li, J., (2016). Formulation of a novel soybean protein - based wood adhesive with desired water resistance and technological applicability. Journal of Applied Polymer Science, 133(27), 564–568.

11. Fulghum, J. E., (1999). Recent developments in high energy and spatial resolution analysis of polymers by XPS. Journal of Electron Spectroscopy and Related Phenomena, 100(1–3), 331–355.
12. Guo, D., Zeng, Z., Wan, Z., Li, Y., Xi, B., & Wang, C., (2021). A CoN - based OER electrocatalyst capable in neutral medium: Atomic layer deposition as rational strategy for fabrication. Advanced Functional Materials, 31(24), 2101324.
13. Hidayat, W., Aprilliana, N., Asmara, S., Bakri, S., Hidayati, S., Banuwa, I. S., & Iswanto, A. H., (2022). Performance of eco - friendly particleboard from agro - industrial residues bonded with formaldehyde - free natural rubber latex adhesive for interior applications. Polymer Composites, 43(4), 2222–2233.
14. Imam, S. H., Bilbao-Sainz, C., Chiou, B. S., Glenn, G. M., & Orts, W. J., (2013). Biobased adhesives, gums, emulsions, and binders: Current trends and future prospects. Journal of Adhesion Science and Technology, 27(18, 19), 1972–1997.
15. Imtiaz, A., Othman, M. H. D., Jilani, A., Khan, I. U., Kamaludin, R., Iqbal, J., & Al-Sehemi, A. G., (2022). Challenges, opportunities, and future directions of membrane technology for natural gas purification: A critical review. Membranes, 12(7), 646.
16. Karana, E., Barati, B., Rognoli, V., & Zeeuw, V. D. L. A., (2015). Material driven design (MDD): A method to design for material experiences. International Journal of Design, 9(2), 35–54.
17. Luo, S., & Van, O. W. J., (2002). Surface modification of textile fibers for improvement of adhesion to polymeric matrices: A review. Journal of Adhesion Science and Technology, 16(13), 1715–1735.
18. Mabrouk, A. B., Dufresne, A., & Boufi, S., (2020). Cellulose nanocrystal as ecofriendly stabilizer for emulsion polymerization and its application for waterborne adhesive. Carbohydrate Polymers, 229, 115504.
19. Mapari, S., Mestry, S., & Mhaske, S. T., (2021). Developments in pressure-sensitive adhesives: A review. Polymer Bulletin, 78, 4075–4108.
20. Momayez, A., Ehsani, M. R., Ramezanianpour, A. A., & Rajaie, H., (2005). Comparison of methods for evaluating bond strength between concrete substrate and repair materials. Cement and Concrete Research, 35(4), 748–757.
21. Morgan, D. R., (1996). Compatibility of concrete repair materials and systems. Construction and Building Materials, 10(1), 57–67.
22. Ohama, Y., (1987). Principle of latex modification and some typical properties of latex-modified mortars and concretes adhesion; binders

(materials); bond (paste to aggregate); carbonation; chlorides; curing; diffusion. Materials Journal, 84(6), 511–518.

23. Patnaik, P. K., Swain, P. T. R., Mishra, S. K., Purohit, A., & Biswas, S., (2020). Recent developments on characterization of needle-punched nonwoven fabric reinforced polymer composites–A review. Materials Today: Proceedings, 26, 466–470.
24. Provencher, J. F., Bond, A. L., Avery-Gomm, S., Borrelle, S. B., Rebolledo, E. L. B., Hammer, S., & Van, F. J. A., (2017). Quantifying ingested debris in marine megafauna: A review and recommendations for standardization. Analytical Methods, 9(9), 1454–1469.
25. Remanan, S., Sharma, M., Bose, S., & Das, N. C., (2018). Recent advances in preparation of porous polymeric membranes by unique techniques and mitigation of fouling through surface modification. ChemistrySelect, 3(2), 609–633.
26. Ryan, C., Ross, S., Davey, P., Duncan, E. M., Francis, J. J., Fielding, S., & Bond, C., (2014). Prevalence and causes of prescribing errors: The PRescribing outcomes for trainee doctors engaged in clinical training (PROTECT) study. PloS One, 9(1), e79802.
27. Saqr, M., Fors, U., Tedre, M., & Nouri, J., (2018). How social network analysis can be used to monitor online collaborative learning and guide an informed intervention. PloS One, 13(3), e0194777.
28. Takeyama, T., & Matsui, J., (1969). Recent developments with tire cords and cord-to-rubber bonding. Rubber Chemistry and Technology, 42(1), 159–256.
29. Verd, J., Abadal, G., Teva, J., Gaudó, M. V., Uranga, A., Borrisé, X., & Barniol, N., (2005). Design, fabrication, and characterization of a submicroelectromechanical resonator with monolithically integrated CMOS readout circuit. Journal of Microelectromechanical Systems, 14(3), 508–519.
30. Xu, L. C., & Logan, B. E., (2005). Interaction forces between colloids and protein-coated surfaces measured using an atomic force microscope. Environmental Science & Technology, 39(10), 3592–3600.
31. Zatirakha, A. V., Smolenkov, A. D., & Shpigun, O. A., (2016). Preparation and chromatographic performance of polymer-based anion exchangers for ion chromatography: A review. Analytica Chimica Acta, 904, 33–50.
32. Zhao, L. F., Liu, Y., Xu, Z. D., Zhang, Y. Z., Zhao, F., & Zhang, S. B., (2011). State of research and trends in development of wood adhesives. Forestry Studies in China, 13, 321–326.

INDEX